C. Faure   Ph. Merloz

Transfixation

Atlas of Anatomical Sections
for the External Fixation of Limbs

Translated by J. E. Robb

With 127 Figures, Mostly in Colour

Springer-Verlag
Berlin Heidelberg New York
London Paris Tokyo

Professor Claude Faure, M. D.
Professor of Anatomy, Orthopaedic Surgeon
University of Grenoble

Doctor Philippe Merloz, M. D.
Orthopaedic Surgeon

Service d'Orthopédie-Traumatologie
Hôpital Nord, BP 217 X, F-38043 Grenoble Cedex

Translator: James E. Robb, F. R. C. S.
14, The Quarry, Alwoodley Park, Leeds LS17 7NH, Great Britain

---

Title of the original French edition: *Transfixion des membres*
© Springer-Verlag Berlin Heidelberg 1987
ISBN-13: 978-3-642-71621-8   e-ISBN-13: 978-3-642-71619-5
DOI: 10.1007/978-3-642-71619-5

Library of Congress Cataloging-in-Publication Data. Faure, C. (Claude), 1942-.
Transfixation : atlas of anatomical sections for the external fixation of limbs.
Translation of: Transfixion des membres.
Includes bibliographies and index.
1. Extremities (Anatomy) – Atlases.   2. Microtomy – Atlases.   3. External skeletal
fixation (Surgery) – Atlases.   I. Merloz, Ph. (Philippe), 1948-.   II. Title. [DNLM:
1. Bone and Bones – injuries – atlases.   2. Extremities – injuries – atlases.   3. Fracture
Fixation – atlases.   4. Orthopedic Fixation Devices – atlases.   WE 17 F265t]
QM548.F3813   1987   611'.98   86-31359

© Springer-Verlag Berlin Heidelberg 1987
Softcover reprint of the hardcover 1st edition 1987

Reproduction of the figures: Gustav Dreher, Württembergische Graphische Kunst-
anstalt GmbH, Stuttgart
Typesetting, Printing and Binding: Appl, Wemding. 2124/3140-543210

Professor François Calas
in memoriam

As used in this volume, the term "transfixation" rather than "transfixion" has been used to convey the notion of piercing through the limb completely in conjunction with a technique of external fixation.

James E. Robb

Foreword

The past 25 years have seen a progressive improvement in external fixation techniques in terms of patient acceptability, ease of application and apparatus stability. The recent awareness in Europe of the method, application and excellent results of the technique pioneered by G. A. Ilizarov has caused a rapid increase here in the number of surgeons who have become "banderilleros". In particular, external fixators can be used in the treatment of difficult clinical conditions, such as severe fractures and their sequelae, as well as for the correction of congenital deformities of the peripheral skeleton.

As one of the first in France to use the Ilizarov method, I became increasingly aware that we needed an updating of anatomical knowledge oriented towards this technique, which uses multiple fixation pins. Without this anatomical information, there is real danger of damaging neuro-vascular structures or entering a joint in the course of assembling the fixator. Traditional anatomical cross-sections are often inadequate or too schematic to serve as a safe guide for the surgeon using this particular technique. This is the reason why my colleagues Claude Faure and Philippe Merloz have produced and analysed a series of anatomical cross-sections of the limbs to serve this particular need.

Colour photographs of the cross-sections show the size and disposition of the neuro-vascular bundles and bone, which are often found in unexpected areas. Each photograph is accompanied by a diagram, shown on the opposite page, which indicates areas related to simple landmarks where transfixation is either safe or dangerous. Cutaneous zones related to the safe areas are also depicted and carefully analysed. Computerised tomographic scans of normal anatomical structures at relevant levels complete the atlas and provide comparisons with aberrant scans obtained in clinical practice.

Claude Faure and Philippe Merloz have provided us with an extremely useful guide. It has been a great pleasure to see the fruits of their work. This atlas is a fine example of their excellent collaboration.

Grenoble, March 1987   Jean Butel

Preface

The decision to produce an atlas of cross-sections of the limbs has been prompted by a growing interest in the use of external fixation devices in the treatment of axial deformities, limb length inequality and fractures. Recently in Europe, there has been a trend favouring the use of devices with fine pins that transfix not only the bone but the entire limb as well, rather than the more traditional unilateral assembly which traverses the bone, but not the whole limb. It is with regard to the former type of system that this atlas has been produced, though it is still of value to those who use the simpler assembly.

Traditionally in the teaching of topographical anatomy, fairly limited use is made of anatomical cross-sections, and we feel that this atlas would be of value to those requiring detailed knowledge of limb anatomy, such as orthopaedic surgeons and physiotherapists. In addition, anatomical drawings often underestimate the size of neurovascular structures, and the position of the bony elements is often depicted as being more central than it is in reality. The surgeon using transfixation pins for the assembly of an external fixator needs to bear in mind safe areas and structures at risk from pin penetration. The originality of the atlas lies in this concept of safe and dangerous areas and gives a guide to cutaneous zones where pin penetration is either safe or hazardous. By studying the cross-sections of the limbs, the surgeon can also choose sites for pin placement so as to minimize muscle damage and consequent loss of joint mobility. The atlas consists of photographs of anatomical cross-sections of normal limbs and commentaries, supplemented by illustrations indicating the safe and dangerous areas. Computerised tomographic scans of normal limbs have also been provided to serve as comparisons with scans obtained in clinical practice.

We would like to point out that the atlas is not intended as a manual of osseous transfixation, or transfixing external fixation, but rather as a guide to the surgeon in the care of his patient, enabling him to choose optimal sites for pin placement.

This atlas could not have been completed without the helpful co-operation of Professor J. Butel and the surgeons working in the University Department of Orthopaedics and Traumatology of Grenoble, as well as Professor Y. Bouchet of the Department of Anatomy in the Faculty of Medicine of Grenoble.

The computerised tomographic scans were produced by the Department of Radiology of the *Centre Hospitalo-Universitaire de Grenoble* under the supervision of Professor M. Coulomb.

X   Preface

We thank Dr. F. Farizon for his help in the preparation of the cross-
sections. Mr. J. E. Robb, FRCS, was responsible for the English
translation of this work. We would particularly like to thank him
for his devotion to his task and for his helpful comments.

Grenoble, March 1987   C. Faure and Ph. Merloz

Contents

Preparation and Analysis of the Cross-sections

An adult cadaver without limb deformity or rotational malalignment of the limbs was used in this study of cutaneous zones for transfixing external fixation. In preparing the cadaver, the following steps were taken:

1. Positioning of the Subject

The limbs were placed in extension to reproduce positions commonly used in surgery. The hip was placed at 15°–20° of abduction with neutral rotation. The upper limb was placed at 45° of abduction with the forearm supinated.

2. Landmarks and Reference Lines

It was necessary to use landmarks and reference lines that are easily determined, since it is by these means that safe zones for external fixation are localised. Once the bony landmarks had been identified, three sets of longitudinal reference lines were made, using an indelible marker: ventral (nearly median), lateral and medial.

3. Deep Freezing

Freezing is necessary before the sections can be made and is preceded by an intra-arterial infusion of 10% formaldehyde solution. This preserving solution prevents deformation of the sections once thawed. The limbs were carefully positioned during freezing to avoid any pressure on the dorsal aspect, thereby preventing any subsequent deformation of the sections. The cadaver was then frozen to −30° Celsius.

4. Axial Landmarks

The levels at which the sections were to be made were marked on the skin after radiographic localisation with the help of a metallic grid fixed to the limb.

5. Cutting the Sections

Sections were cut perpendicular to the longitudinal axis of the limbs. Each section, measuring 13–15 mm in thickness, was then left to thaw, after which it was brushed and then stored in 10% formaldehyde solution.

6. Analysis of the Sections and Illustrations

The reference lines were located on each section and marked with a pin. The frontal and sagittal planes were also determined in this manner. Photographs were then taken of the distal surface of the section. The choice of this surface permits subsequent comparisons with computerised tomographic scans. The skeletal elements are usually found neither at the intersection of the frontal and sagittal planes nor in the centre of the section. The middle of the medullary canal, being the target of the pins used for fixation, was marked along with the superficial and deep vessels, nerves, tendons, muscle bellies and joints. Both the areas that are safe for transfixing external fixation and the dangerous zones were then determined by pivoting straight lines representing a transfixing pin around the medullary cavity. The cutaneous areas corresponding to these safe and dangerous zones were also localised. Normally, four zones, alternately safe and dangerous, can be seen in the circumference of the section.

Narrow areas where external fixation is possible are found on either side of the safe zones. These narrow sections are generally limited by anatomical structures that lie close together. External fixation is possible in these regions, though with much less safety, and for this reason, we do not recommend them.

In the forearm, the radius and ulna can be transfixed either together or separately. Here, the safe zones have been indicated for both combined fixation of the two bones and individual fixation of the radius and the ulna.

7. Longitudinal Analysis

Using the information from the cross-sections, it is possible to determine cutaneous zones in the longitudinal axis where external fixation is either safe or dangerous. In clinical practice, these zones can also be mapped with reference to bony landmarks and the axial reference lines.

8. Computerised Tomographic Scans

The same method of marking the limbs for the anatomical sections was used to determine the levels for the computerised tomographic scans.

A. Cross-sections of the Shoulder and Arm

Levels of the Cross-sections

These cross-sections are arranged into four groups corresponding to the areas commonly used for external fixation.

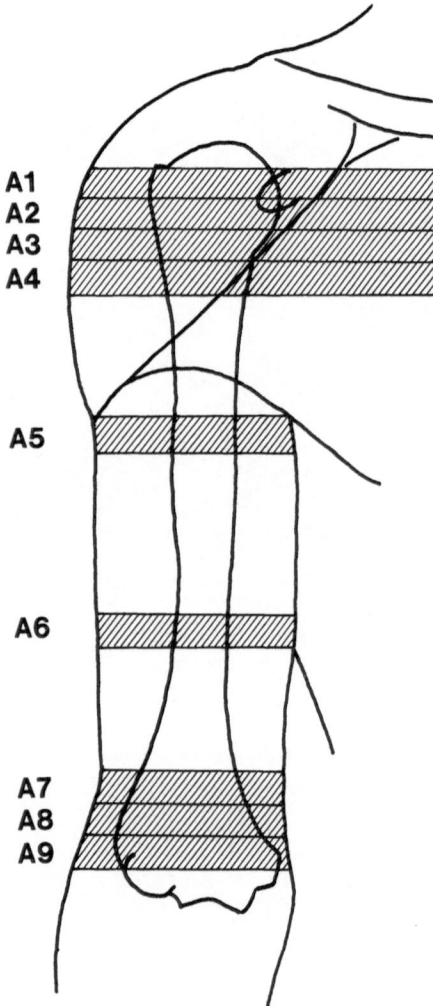

Proximal Epiphysis and Metaphysis (A 1–A 4)

The four sections have been taken at the level of the humeral head, the tuberosities and the surgical neck of the humerus.

Proximal Diaphysis (A 5)

This section has been taken at the junction of the proximal and middle thirds of the humeral diaphysis, above the insertion of the deltoid muscle.

Distal Diaphysis (A 6)

This section has been taken at the junction of middle and distal thirds of the humeral diaphysis, close to the origin of the brachio-radialis muscle.

Distal Metaphysis and Epiphysis (A 7–A 9)

These three sections have been taken at the distal expansion of the humerus. The last includes the medial and lateral epicondyles, the trochlea and the olecranon.

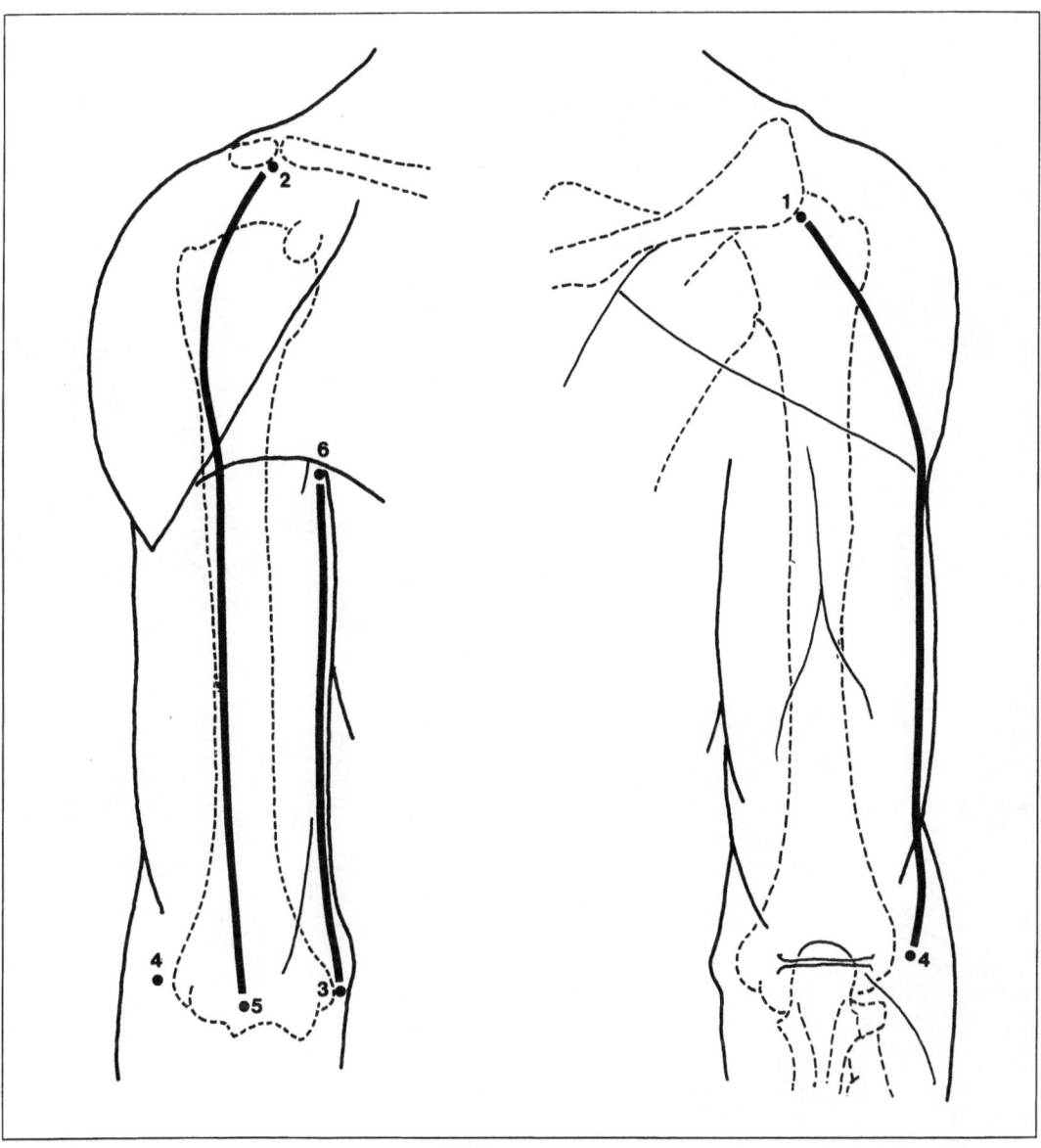

Landmarks

1 Angle of the acromion
2 Acromio-clavicular joint
3 Medial epicondyle
4 Lateral epicondyle
5 Middle of the transverse skin crease of the elbow (line 3–4)
6 Area where the brachial artery pulse is palpable at the base of the arm

Reference lines

Ventral line: line 2–5
Lateral line: line 1–4
Medial line: line 3–6

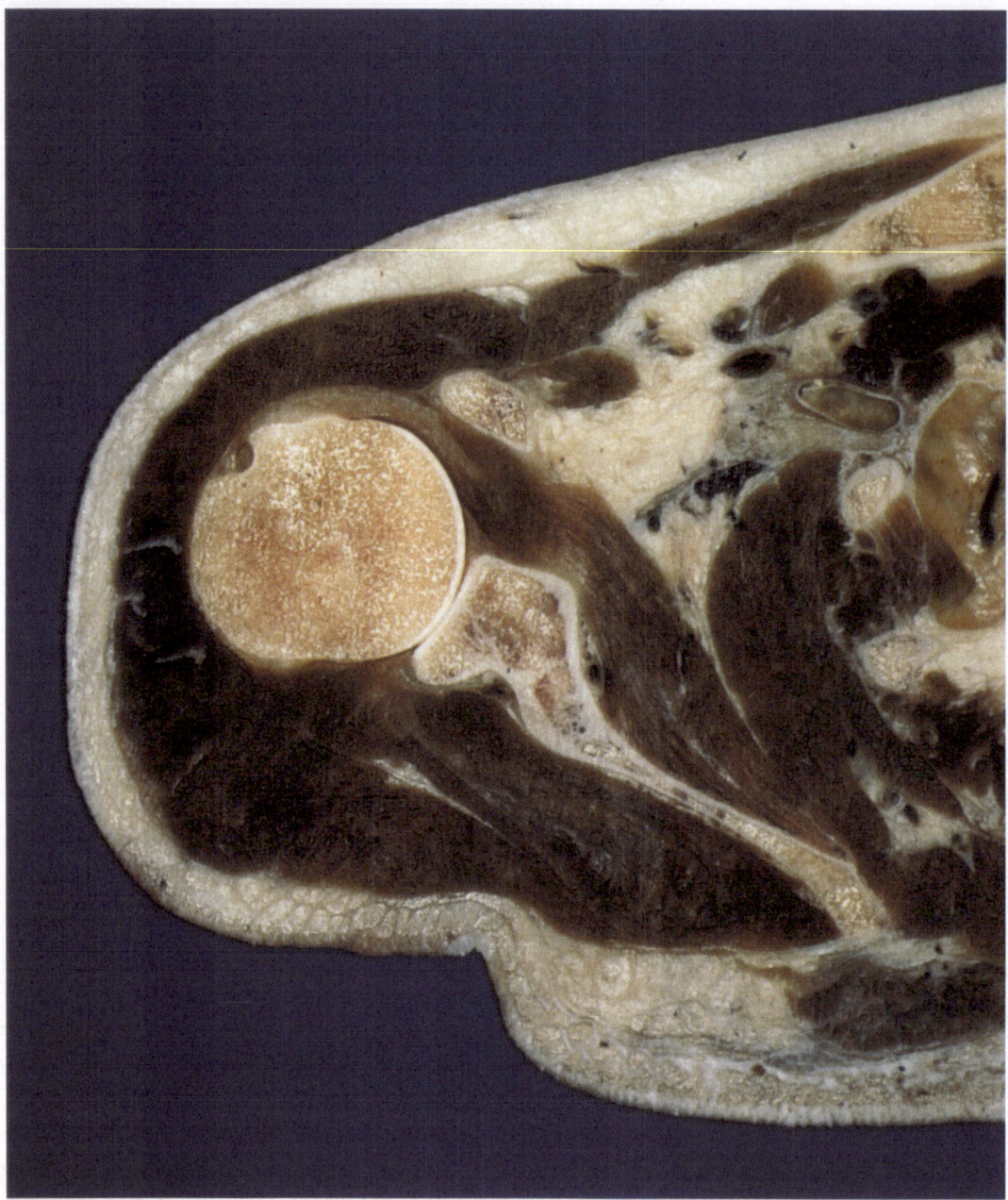

Cross-section A 1

Bone. The section shows the scapulo-humeral joint, the greater and the lesser tuberosities of the humerus and the coracoid process.

Vessels and Nerves. The axillary vessels and the medial, lateral and dorsal cords of the brachial plexus are at some distance from the humerus.

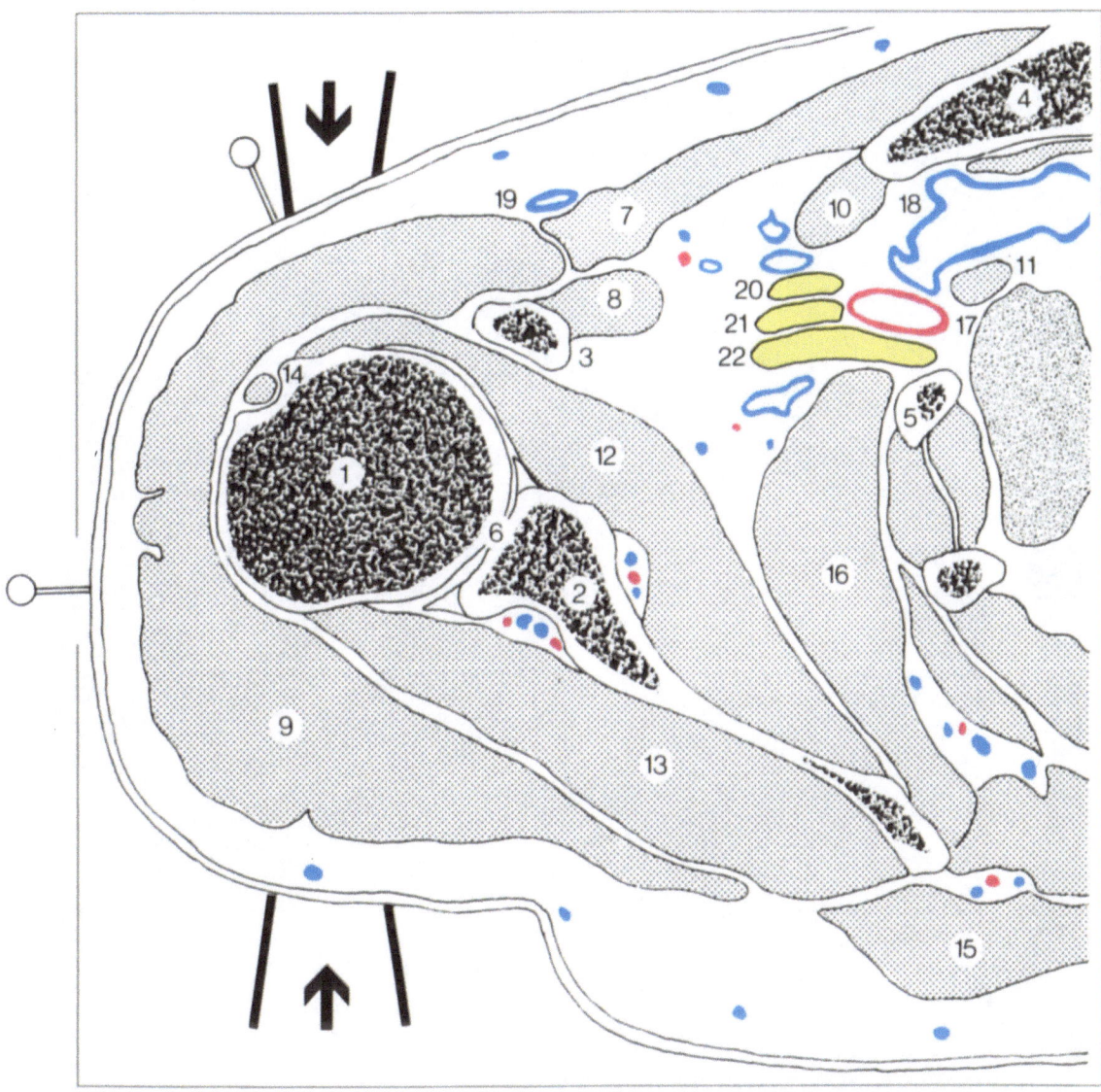

1 Humeral head	9 Deltoid	16 Serratus anterior
2 Scapula (neck)	10 Subclavius	17 Axillary artery
3 Scapula (coracoid process)	11 Scalenus anterior	18 Axillary vein
4 Clavicle	12 Subscapularis	19 Cephalic vein
5 First rib	13 Supraspinatus	20 Brachial plexus (medial cord)
6 Scapulo-humeral joint	14 Biceps brachii (long head)	21 Brachial plexus (lateral cord)
7 Pectoralis major	15 Trapezius	22 Brachial plexus (dorsal cord)
8 Pectoralis minor		

Safe Areas. These are ventral and dorsal to the humerus, but the size of the joint reduces their area.

Cutaneous Zones Related to the Safe Areas. These comprise two narrow zones on the ventro-lateral and dorso-lateral surfaces in relation to the reference lines.
- *The ventro-lateral zone* corresponds to the palpable bony prominence of the two tuberosities, which are separated by the bicipital groove.
- *The narrow dorso-lateral zone* lies diametrically opposite and is centred on the junction of the acromial and scapular portions of the deltoid muscle.

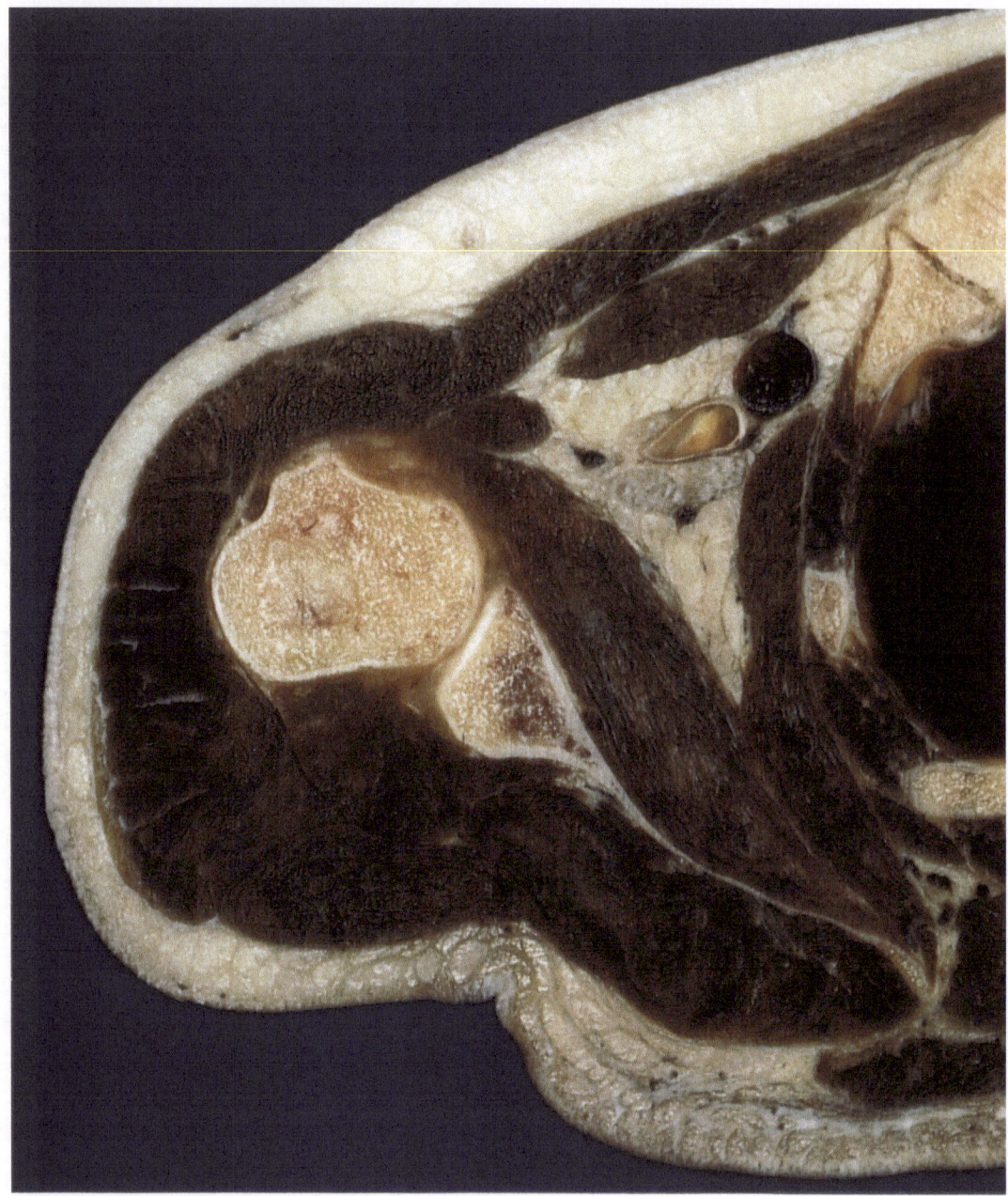

Cross-section A 2

Bone. The section shows the inferior aspect of the humeral head and the lesser tuberosity.

Vessels and Nerves. The axillary vessels and the cords of the brachial plexus are some distance apart.

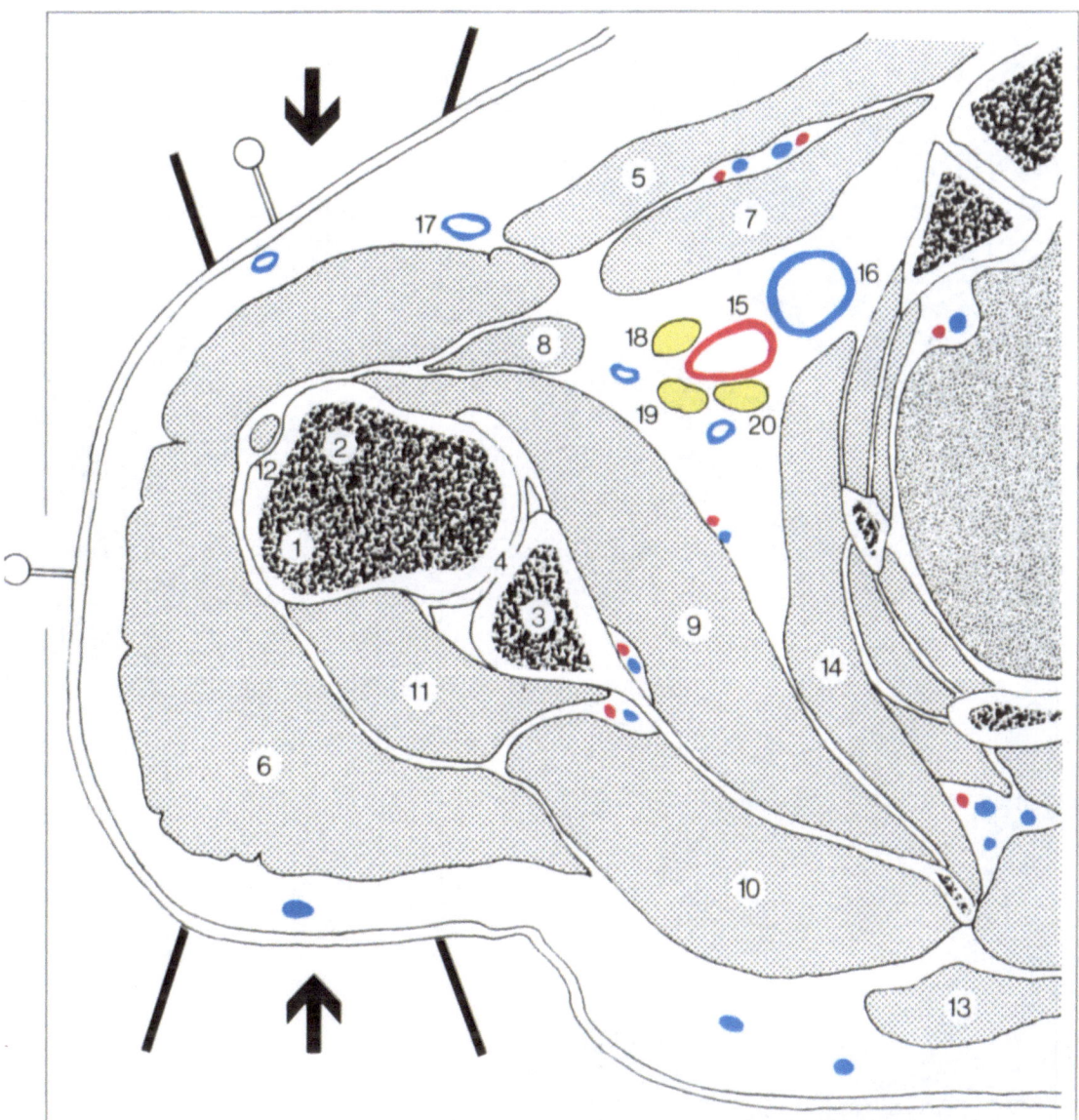

1 Humerus (greater tuberosity)	8 Coracobrachialis +	14 Serratus anterior
2 Humerus (lesser tuberosity)	biceps brachii (short head)	15 Axillary artery
3 Scapula	9 Subscapularis	16 Axillary vein
4 Scapulo-humeral joint	10 Infraspinatus	17 Cephalic vein
5 Pectoralis major	11 Teres minor	18 Brachial plexus (medial cord)
6 Deltoid	12 Biceps brachii (long head)	19 Brachial plexus (lateral cord)
7 Pectoralis minor	13 Trapezius	20 Brachial plexus (dorsal cord)

Safe Areas. These are narrow and are found on the ventro-lateral and dorso-lateral surfaces with respect to the humerus.

Cutaneous Zones Related to the Safe Areas. These comprise two zones on the ventro-lateral and dorso-lateral surfaces in relation to the reference lines.
- *The ventro-lateral zone* is centred on the bony prominence of the lesser tuberosity and corresponds to the lateral half of the clavicular portion of the deltoid muscle.
- *The dorso-lateral zone* corresponds to the lateral half of the scapular portion of the deltoid muscle.

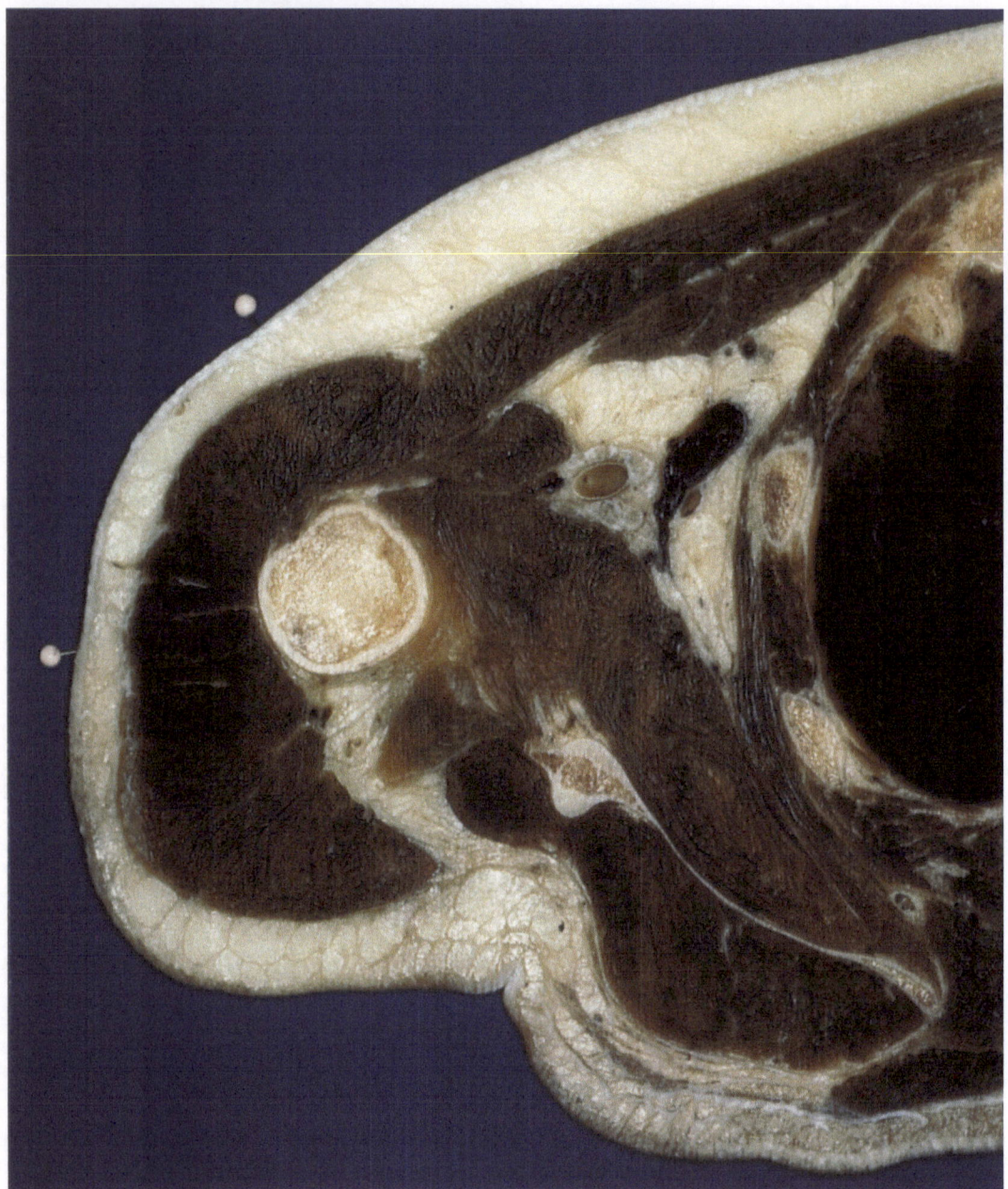

Cross-section A 3

Bone. The section has been taken immediately distal to the scapulo-humeral joint.

Vessels and Nerves. The axillary vessels and the cords of the brachial plexus are some distance apart.

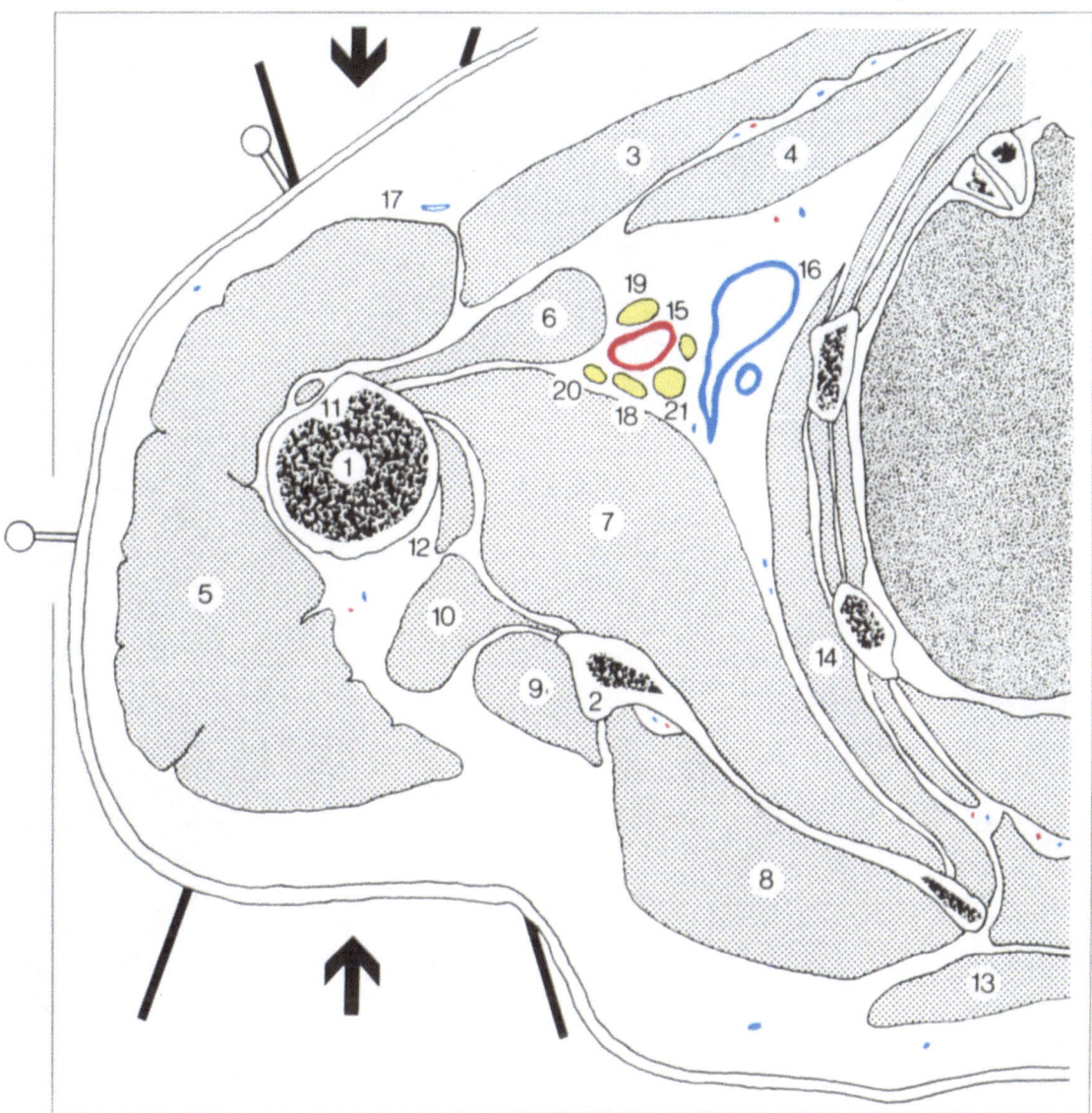

1 Humerus	8 Infraspinatus	15 Axillary artery
2 Scapula	9 Teres minor	16 Axillary vein
3 Pectoralis major	10 Teres major + latissimus dorsi	17 Cephalic vein
4 Pectoralis minor	11 Biceps brachii (long head)	18 Radial nerve
5 Deltoid	12 Triceps brachii (long head)	19 Median nerve
6 Coracobrachialis + biceps brachii	13 Trapezius	20 Axillary nerve
(short head)	14 Serratus anterior	21 Ulnar nerve
7 Subscapularis		

Safe Areas. These are narrow and are found on the ventral and dorsal surfaces in relation to the bone.

Cutaneous Zones Related to the Safe Areas. These comprise two zones on the ventral and dorsal surfaces in relation to the reference lines.

External fixation should avoid the dorsal and ventral boundaries of the axilla.

- *The ventral zone* corresponds to the palpable portion of the humeral metaphysis and lies between the cephalic vein and the ventral reference line.
- *The dorsal zone* corresponds to the dorsal portion of the deltoid muscle.

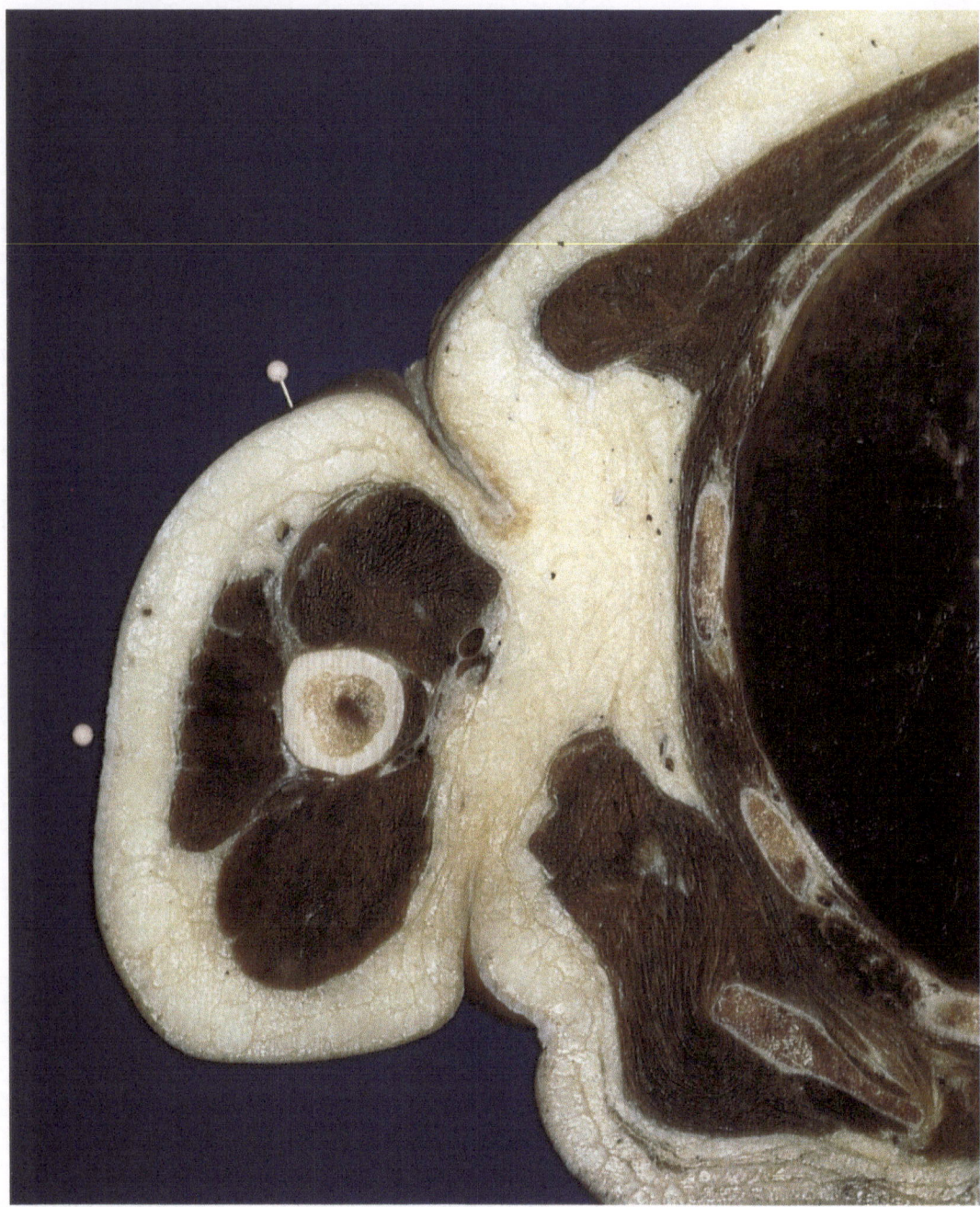

Cross-section A 4

Bone. The section has been taken at the surgical neck of the humerus. The arm is separate from the axilla.

Vessels and Nerves. The vessels and nerves are now closer to the bone and lie medially.

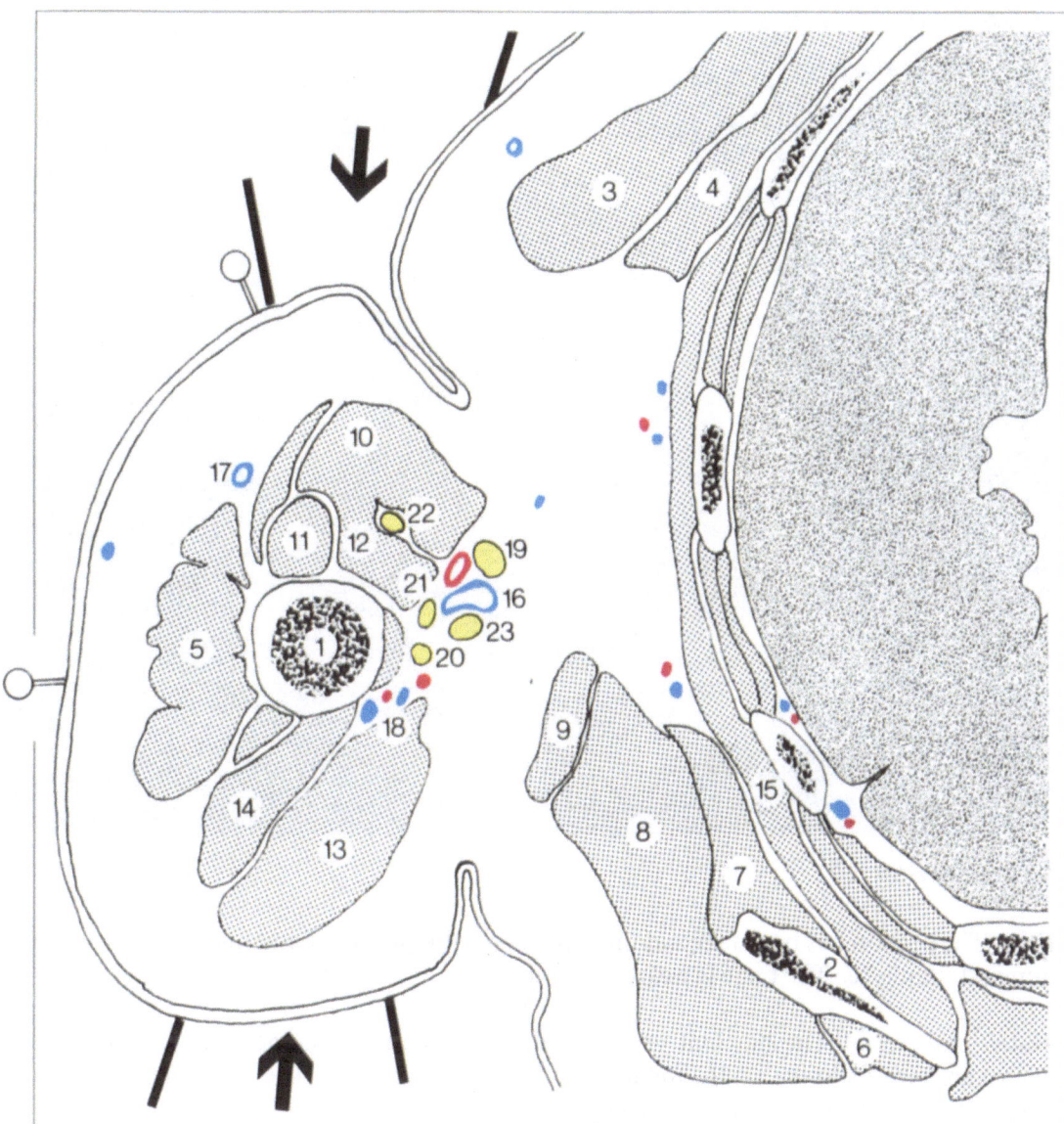

1 Humerus (surgical neck)	9 Latissimus dorsi	17 Cephalic vein
2 Scapula	10 Biceps brachii (short head)	18 Posterior circumflex artery
3 Pectoralis major	11 Biceps brachii (long head)	and vein
4 Pectoralis minor	12 Coracobrachialis	19 Median nerve
5 Deltoid	13 Triceps brachii (long head)	20 Axillary nerve
6 Infraspinatus	14 Triceps brachii (lateral head)	21 Radial nerve
7 Subscapularis	15 Serratus anterior	22 Musculocutaneous nerve
8 Teres major	16 Brachial artery and vein	23 Ulnar nerve

Safe Areas. These are smaller and are found on the ventral and dorsal surfaces in relation to the humerus.

Cutaneous Zones Related to the Safe Areas. These comprise two zones on the ventro-medial and dorsal surfaces in relation to the reference lines.
- *The ventro-medial zone* lies between the axillary fold and the ventral reference line.
- *The dorsal zone* corresponds to the body of the long head of the triceps brachii muscle.

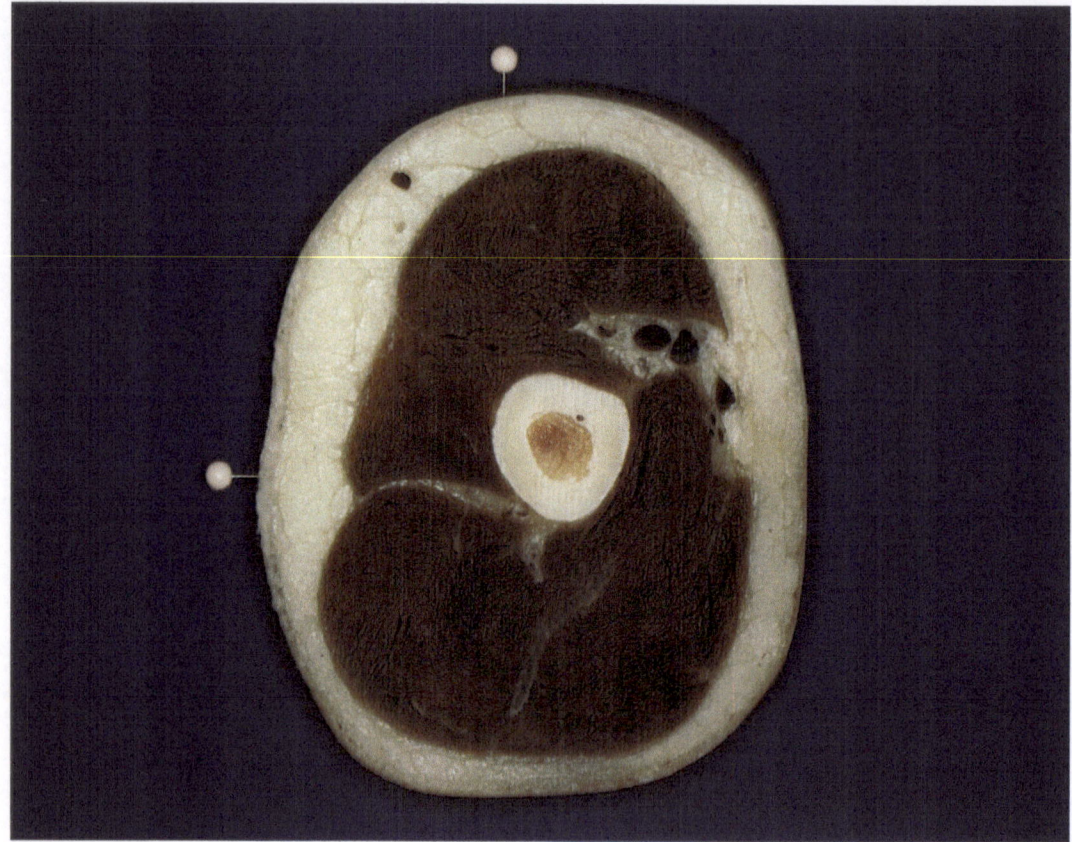

Cross-section A 5

Bone. The section has been taken at the junction of the proximal and middle thirds of the humeral diaphysis. The bone is nearly central.

Vessels and Nerves. The brachial vessels and the median, musculocutaneous and ulnar nerves are located in the ventro-medial quadrant of the section. The radial nerve and the deep vessels of the arm are diametrically opposed.

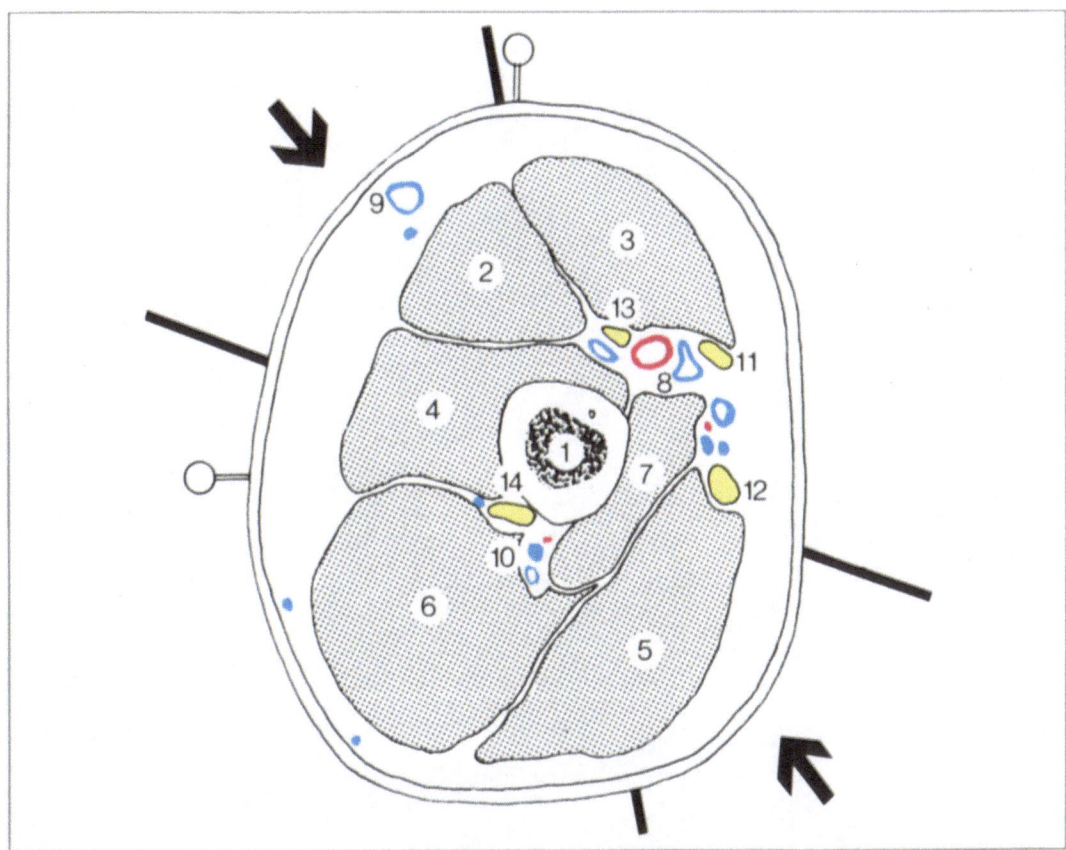

1 Humerus (diaphysis)
2 Biceps brachii (long head)
3 Biceps brachii (short head)
4 Brachialis
5 Triceps brachii (long head)
6 Triceps brachii (lateral head)
7 Triceps brachii (medial head)
8 Brachial artery and vein
9 Cephalic vein
10 Profunda brachii artery and vein
11 Median nerve
12 Ulnar nerve
13 Musculocutaneous nerve
14 Radial nerve

Safe Areas. These are of moderate size and lie ventro-laterally and dorso-medially in relation to the humerus.

Cutaneous Zones Related to the Safe Areas. These comprise two zones on the ventro-lateral and dorso-medial surfaces in relation to the reference lines.

- *The ventro-lateral zone* is found on the ventral two-thirds of the area between the ventral and lateral reference lines.

- *The dorso-medial zone* corresponds to the middle third of the body of the long head of the triceps brachii muscle.

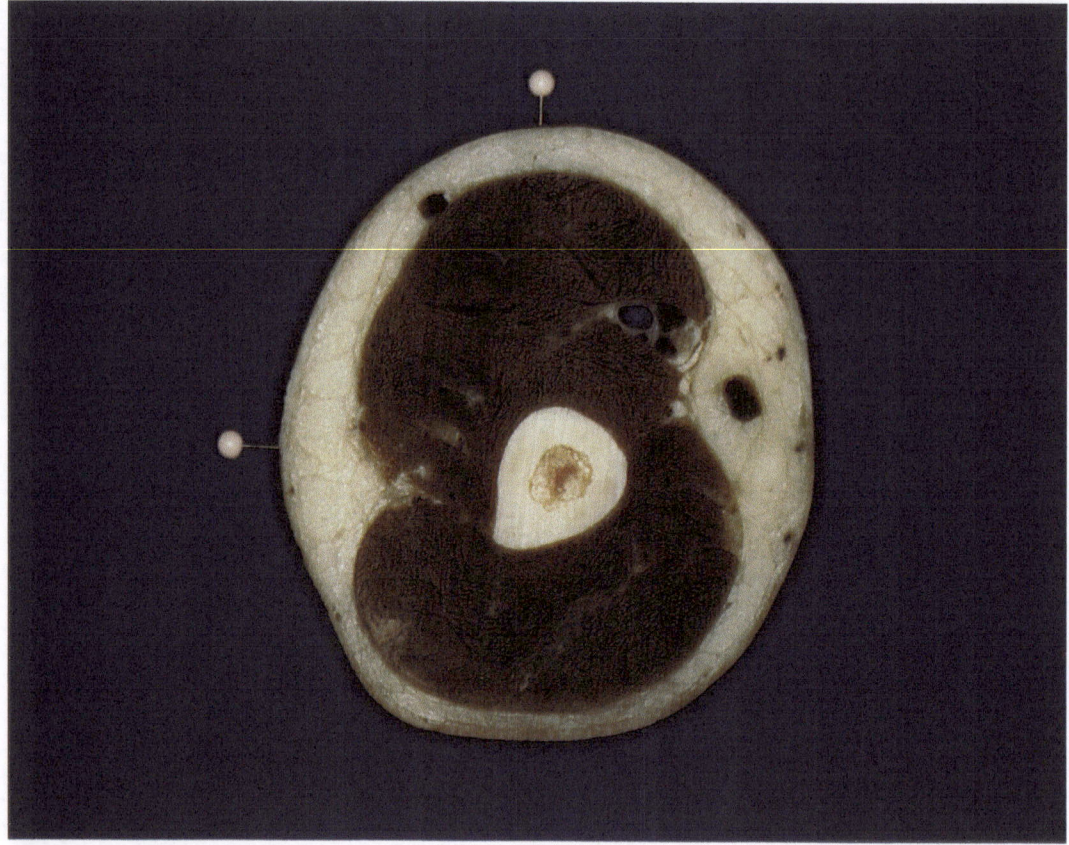

Cross-section A 6

Bone. The section has been taken at the junction of the middle and distal thirds of the humeral diaphysis. The bone is becoming triangular and has a nearly central position.

Vessels and Nerves. The brachial vessels and the musculocutaneous and median nerves lie in the ventro-medial quadrant. The ulnar nerve lies more dorsally and the radial nerve lies between the biceps brachii, brachioradialis and brachialis muscles.

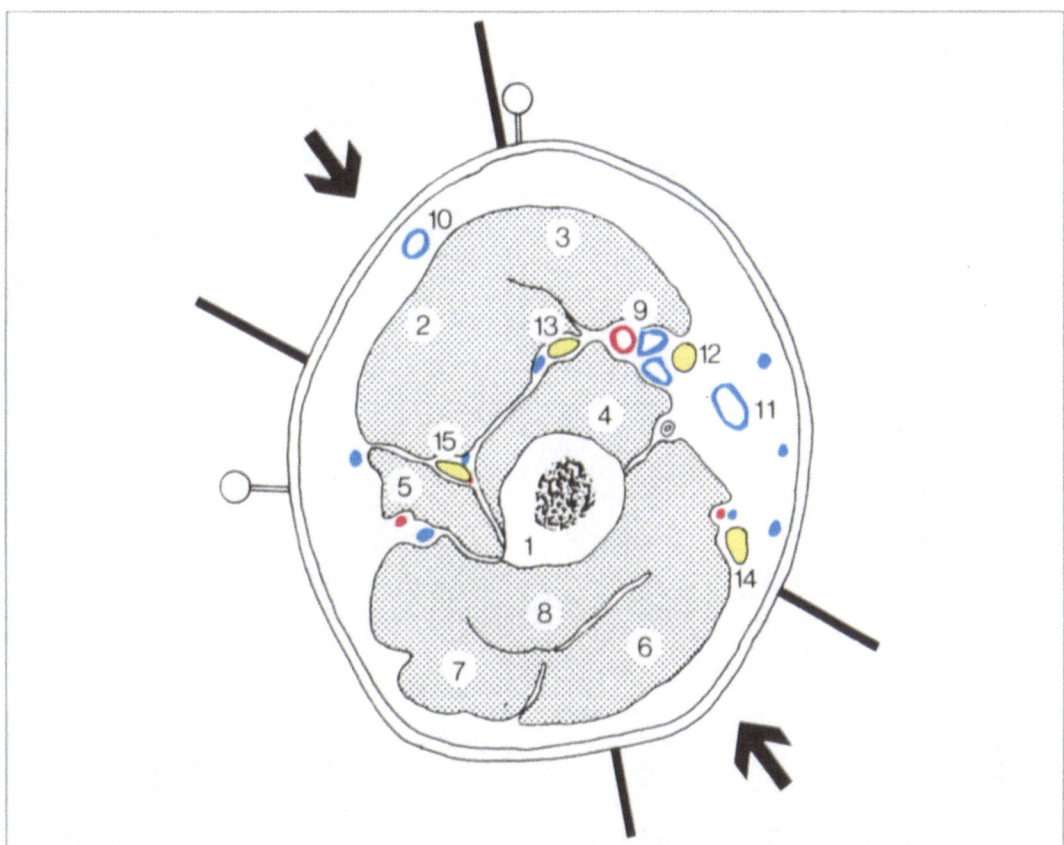

1 Humerus	6 Triceps brachii (long head)	11 Basilic vein
2 Biceps brachii (long head)	7 Triceps brachii (lateral head)	12 Median nerve
3 Biceps brachii (short head)	8 Triceps brachii (medial head)	13 Musculocutaneous nerve
4 Brachialis	9 Brachial artery and vein	14 Ulnar nerve
5 Brachioradialis	10 Cephalic vein	15 Radial nerve

Safe Areas. These are of moderate size and lie on the ventro-lateral and dorso-medial surfaces in relation to the humerus.

Cutaneous Zones Related to the Safe Areas. These comprise two zones on the ventro-lateral and dorso-medial surfaces in relation to the reference lines.

- *The ventro-lateral zone* is found on the ventral two-thirds of the area between the ventral and lateral reference lines.

- *The dorso-medial zone* corresponds to the middle third of the body of the long head of the triceps brachii muscle.

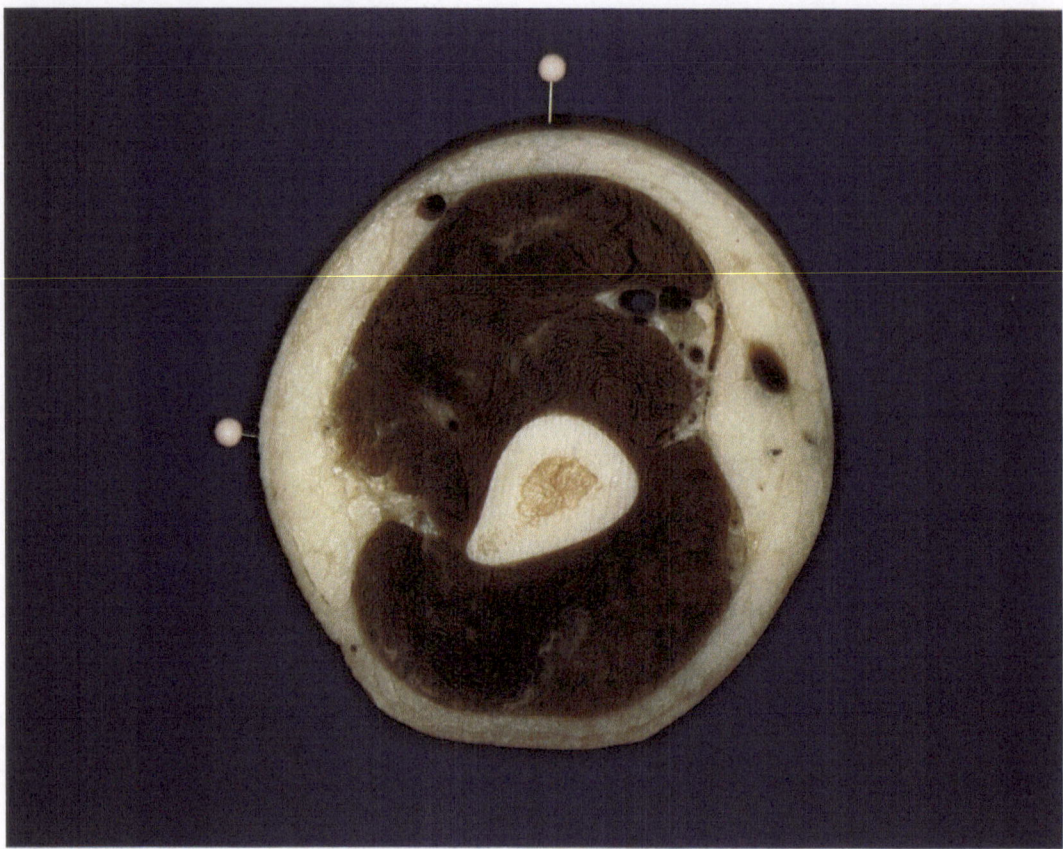

Cross-section A 7

Bone. The section has been taken at the distal metaphysis of the humerus. The bone now lies slightly off-centre and dorsally.

Vessels and Nerves. The brachial vessels and the median nerve lie ventro-medially. The radial nerve lies between the brachialis and brachioradialis muscles, while the ulnar nerve lies dorso-medially. The musculocutaneous nerve now consists of sensory branches only.

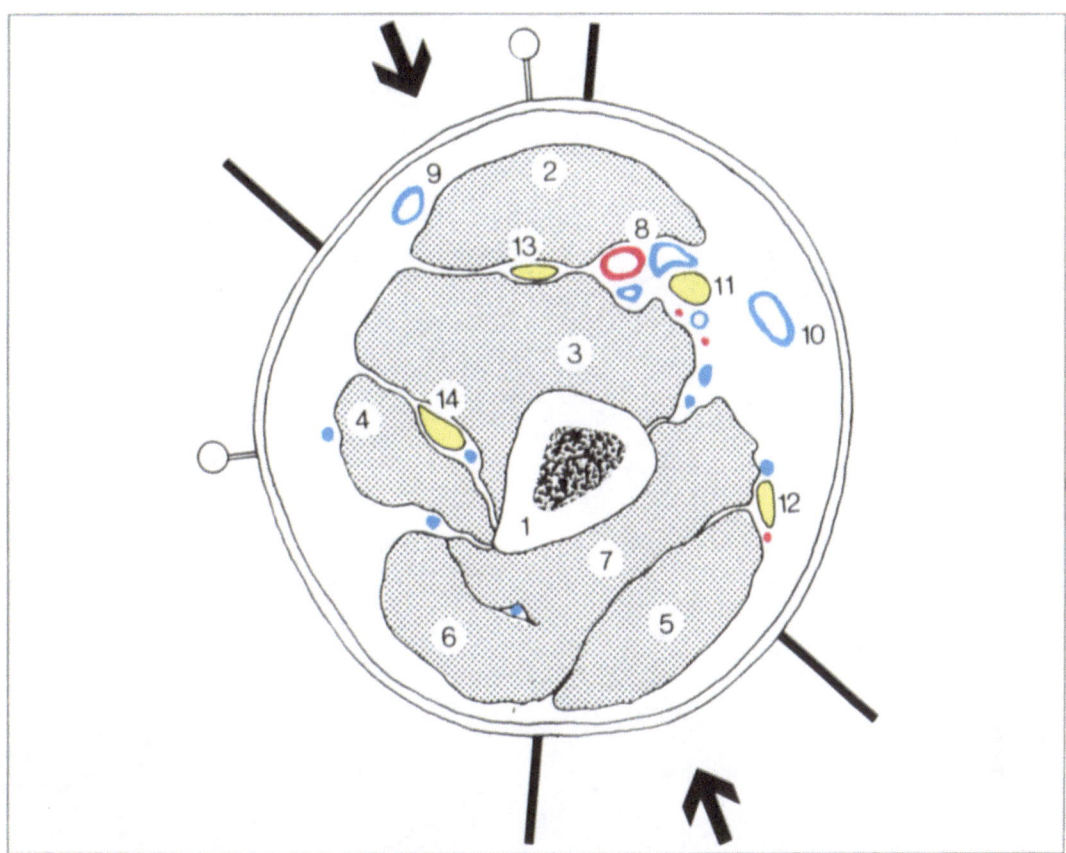

1 Humerus	6 Triceps brachii (lateral head)	11 Median nerve
2 Biceps brachii	7 Triceps brachii (medial head)	12 Ulnar nerve
3 Brachialis	8 Brachial artery and vein	13 Musculocutaneous nerve
4 Brachioradialis	9 Cephalic vein	14 Radial nerve
5 Triceps brachii (long head)	10 Basilic vein	

Safe Areas. These lie on the ventro-lateral and dorso-medial surfaces in relation to the humerus.

Cutaneous Zones Related to the Safe Areas. These comprise two zones on the ventro-lateral and dorso-medial surfaces in relation to the reference lines.

- *The ventro-lateral zone* is found on the ventral two-thirds of the area between the ventral and lateral reference lines.

- *The dorso-medial zone* lies dorsal to the medial border of the distal metaphysis, which is readily palpable here. This corresponds to the dorsal half of the long head of the triceps brachii muscle.

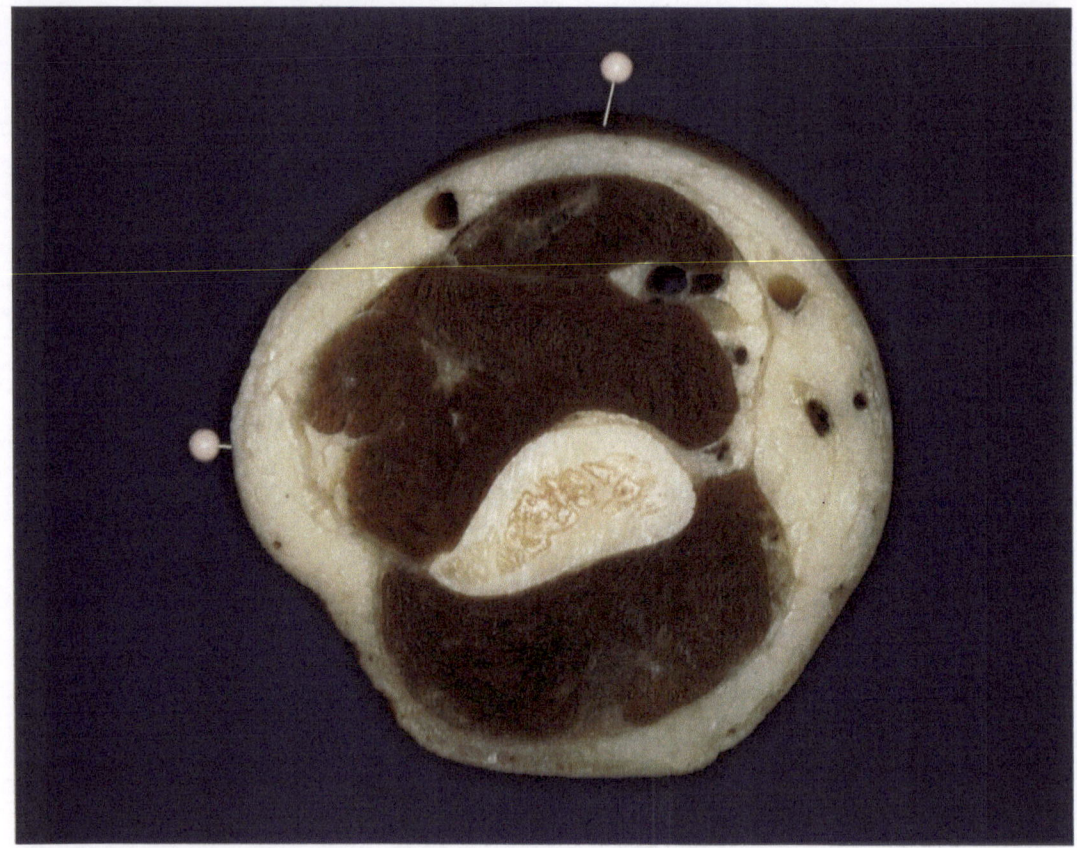

Cross-section A 8

Bone. The section has been taken through the metaphyseal flare of the humerus. The medial and lateral borders are superficial.

Vessels and Nerves. The brachial vessels and the median and radial nerves are ventral. The musculocutaneous nerve has divided into its cutaneous branches. The ulnar nerve is dorsal.

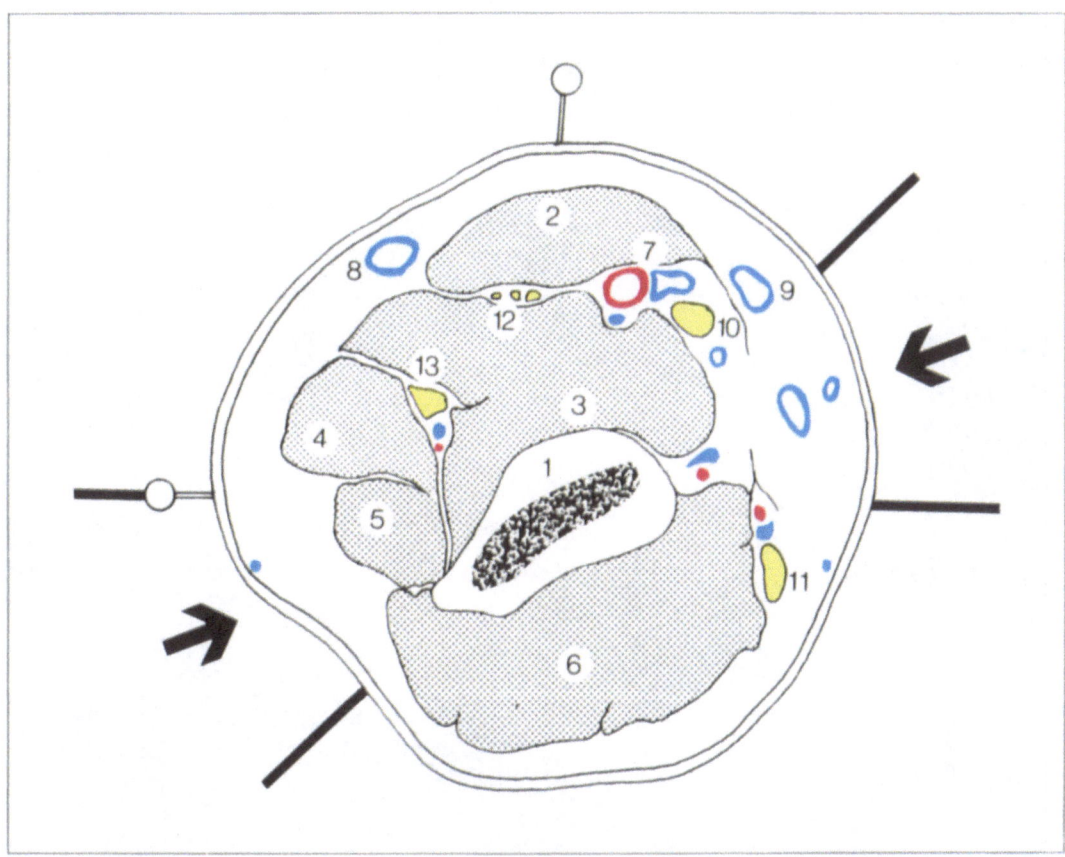

1 Humerus	6 Triceps brachii	10 Median nerve
2 Biceps brachii	7 Brachial artery and vein	11 Ulnar nerve
3 Brachialis	8 Cephalic vein	12 Musculocutaneous nerve
4 Brachioradialis	9 Basilic vein	13 Radial nerve
5 Extensor carpi radialis longus		

Safe Areas. These are reduced and lie on the lateral and ventro-medial surfaces in relation to the humerus.

Cutaneous Zones Related to the Safe Areas. These comprise two zones on the dorso-lateral and ventro-medial surfaces in relation to the reference lines. They are narrow owing to the flattening of the humerus in this region.

- *The dorso-lateral zone* is found between the lateral bony border, which is palpable, and the lateral reference line.

- *The ventro-medial zone* lies between the medial reference line and the course of the median nerve and brachial artery, the latter being palpable here.

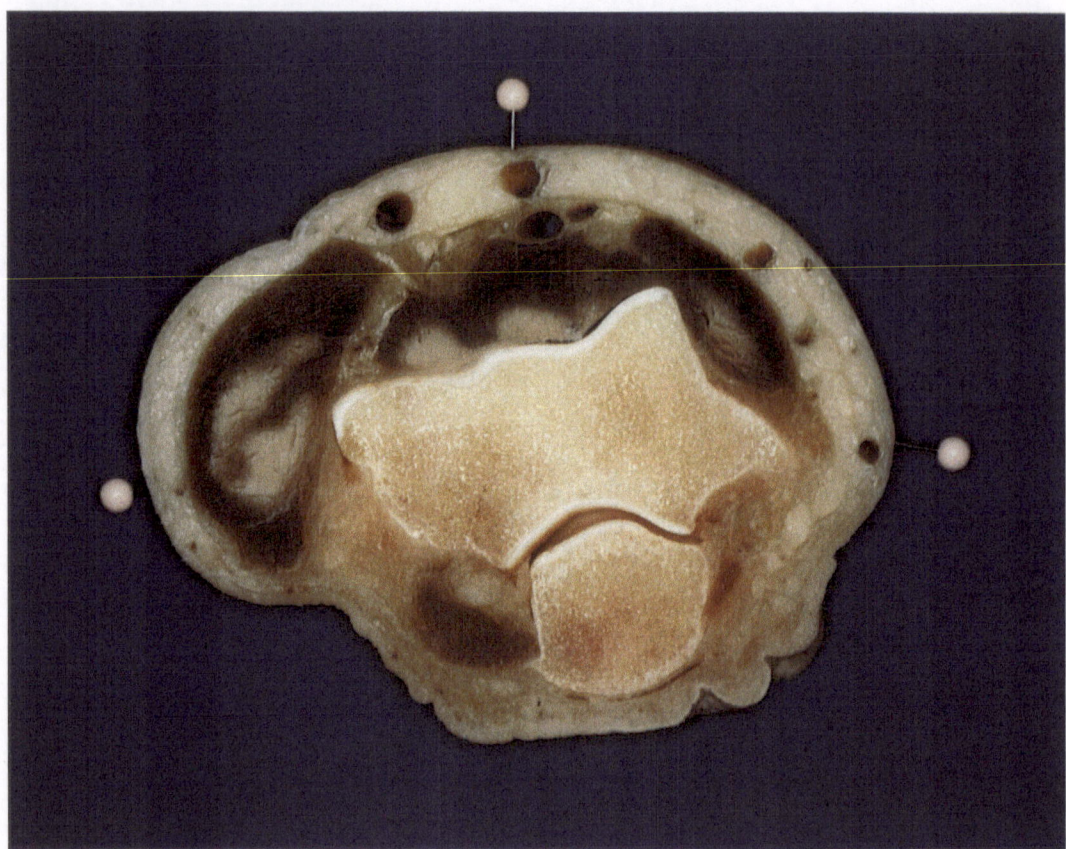

Cross-section A 9

Bone. The section shows the elbow joint, the medial epicondyle, the trochlea and the olecranon.

Vessels and Nerves. The brachial vessels accompanied by the median nerve lie ventrally and close to the joint. The ulnar nerve lies in the medial epicondylar groove.

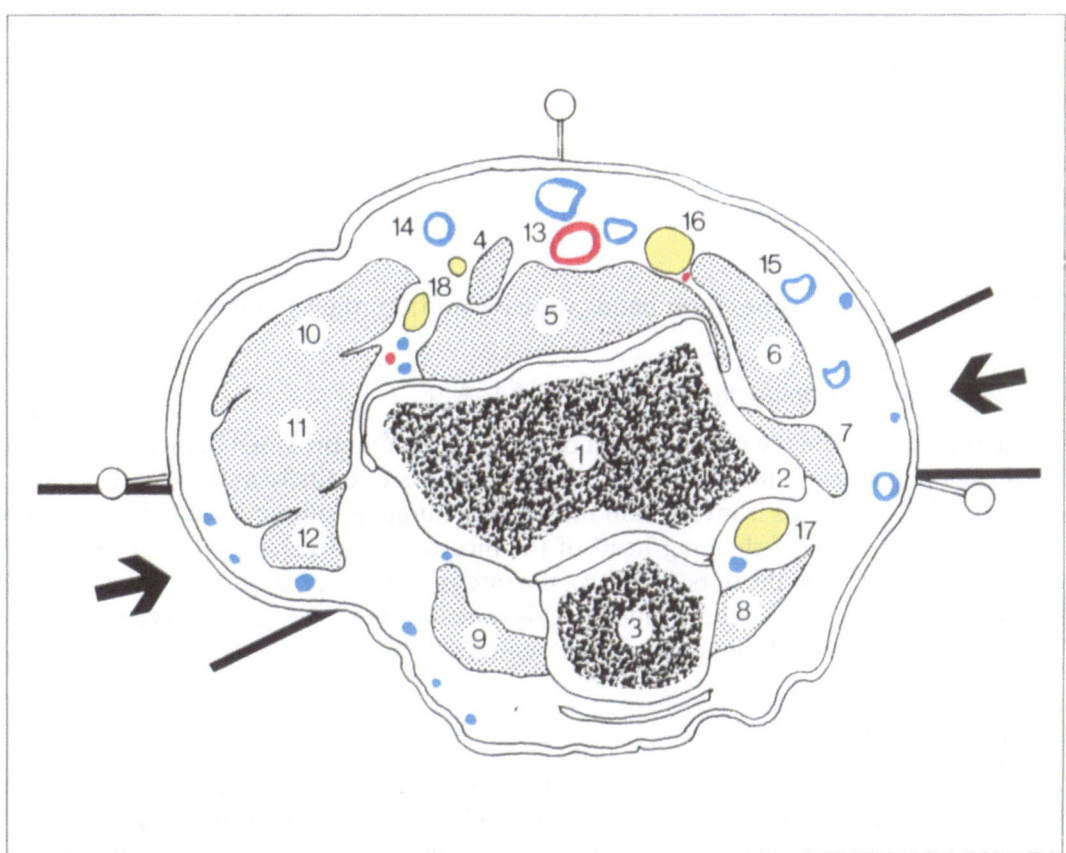

1 Humerus (distal epiphysis)	7 Flexor carpi radialis	13 Brachial artery and vein
2 Humerus (medial epicondyle)	8 Flexor carpi ulnaris	14 Cephalic vein
3 Olecranon	9 Anconeus	15 Basilic vein
4 Biceps (distal tendon)	10 Brachioradialis	16 Median nerve
5 Brachialis	11 Extensor carpi radialis longus	17 Ulnar nerve
6 Pronator teres	12 Extensor carpi radialis brevis	18 Radial nerve

Safe Areas. These are much reduced and are medial and lateral in relation to the humerus.

Cutaneous Zones Related to the Safe Areas. These comprise two zones on the dorso-lateral and medial surfaces in relation to the reference lines.

- *The narrow dorso-lateral zone* consists of the ventral half of the area between the lateral reference line and the lateral border of the olecranon.

- *The medial zone* lies between the prominence of the medial epicondyle and the middle of the body of the pronator teres muscle.

Safe Zones of the Shoulder and Arm

The safe cutaneous zones form bands which wind from the shoulder towards the elbow.

Shoulder and Proximal Humerus (Sections A1–A3)

External fixation on the ventral aspect of the shoulder and proximal humerus is limited to the area palpable around the lesser tuberosity. This is due firstly to the fact that much of the proximal humerus is involved in the scapulo-humeral joint and secondly to the presence of the tendon of the long head of the biceps brachii muscle lying in the bicipital groove. However, on the dorsal surface, the safe zone is based over the medial half of the scapular portion of the deltoid muscle. This zone is larger owing to the eccentric position of the bone ventrally. It lies opposite the ventral safe zone.

Diaphysis (Sections A4–A7)

The arm is separate from the axilla, and the brachial artery and the median and ulnar nerves form a longitudinal medial neurovascular axis in relation to the humerus. The radial and the musculocutaneous nerves, separated by the diaphysis, run obliquely towards the lateral bicipital groove. The musculocutaneous nerve gives branches to the biceps and brachialis muscles early in its course, which lessens the risk of damage by external fixation. The safe zones continue distally from those around the shoulder. The ventral zone enlarges and becomes ventro-lateral in relation to the distal humeral diaphysis, whereas the zone which is dorsal at the level of the axilla becomes dorso-medial. The two zones change direction in same sense as the course of the brachial artery.

Distal Metaphyseal and Epiphyseal Regions (Sections A8, A9)

External fixation is possible only on the two narrow cutaneous surfaces overlying the medial and lateral epicondyles. This is due to the situation of the articular surfaces on the ventral and dorsal aspects of the distal humerus and the position of the vessels and nerves at this level.

The safe zone for external fixation seen on the ventral aspect has been designated the ventral safe zone (band 1) and that seen on the dorsal aspect, the dorsal safe zone (band 2).

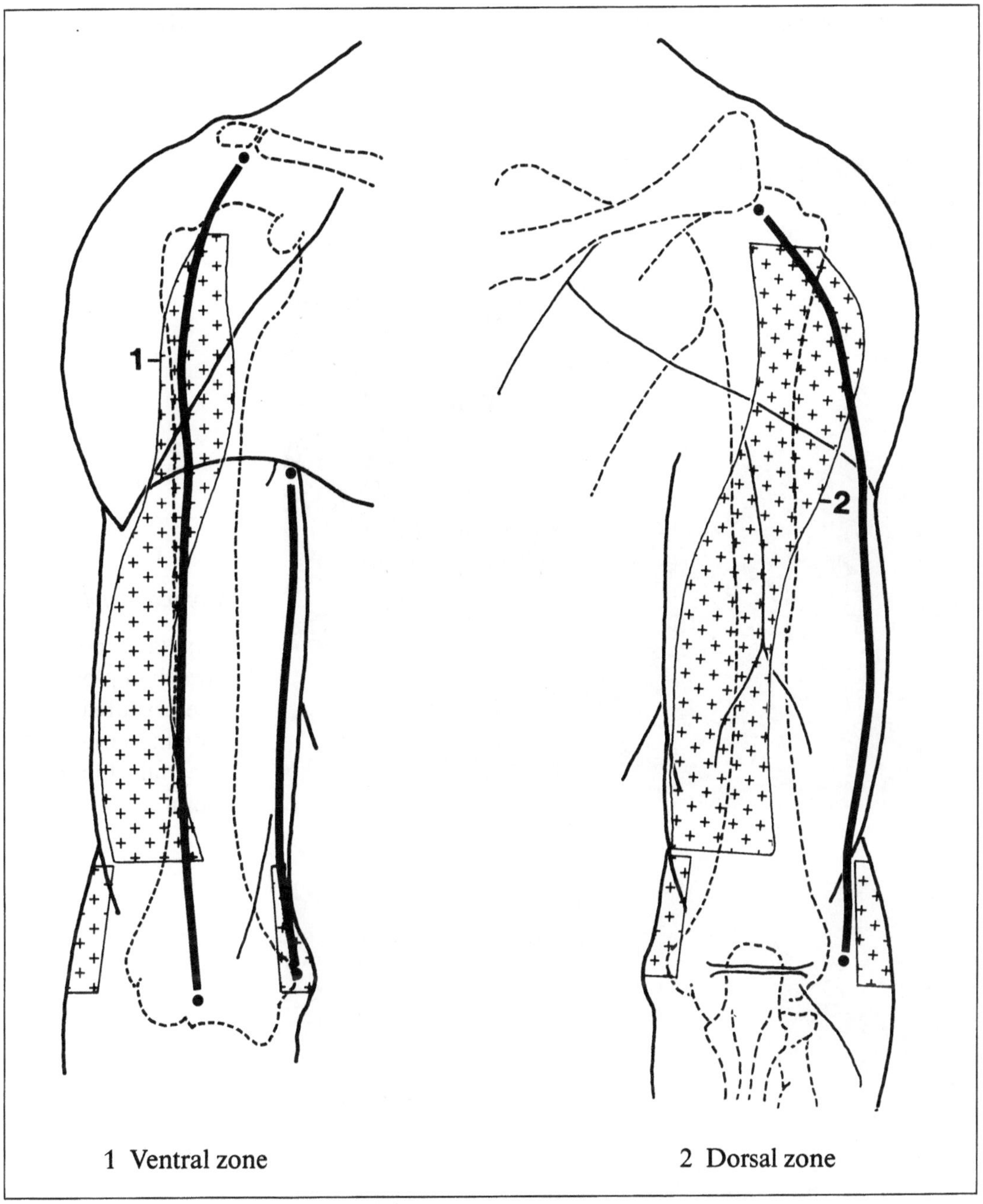

1 Ventral zone   2 Dorsal zone

Computerised Tomographic Scans of the Shoulder and Arm

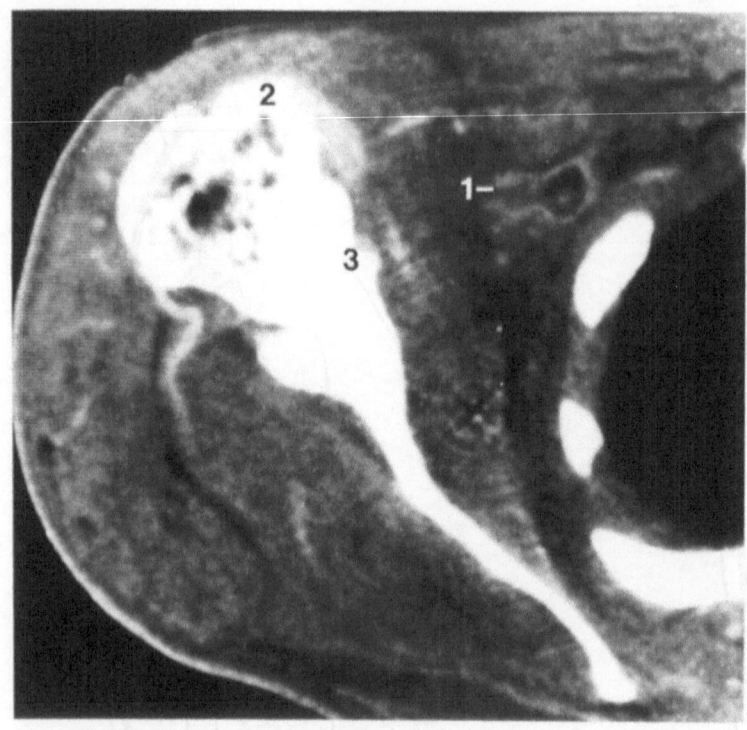

Proximal epiphysis

1 Axillary vessels +
  brachial plexus
2 Lesser tuberosity of
  the humerus
3 Scapulo-humeral
  joint

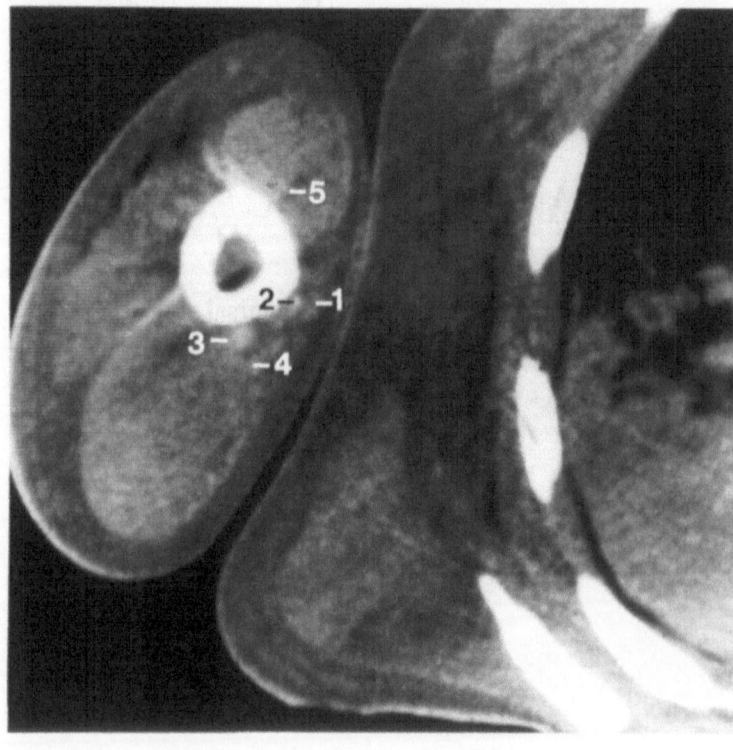

Proximal diaphysis

1 Brachial vessels
2 Median nerve
3 Radial nerve
4 Ulnar nerve
5 Musculocutaneous
  nerve

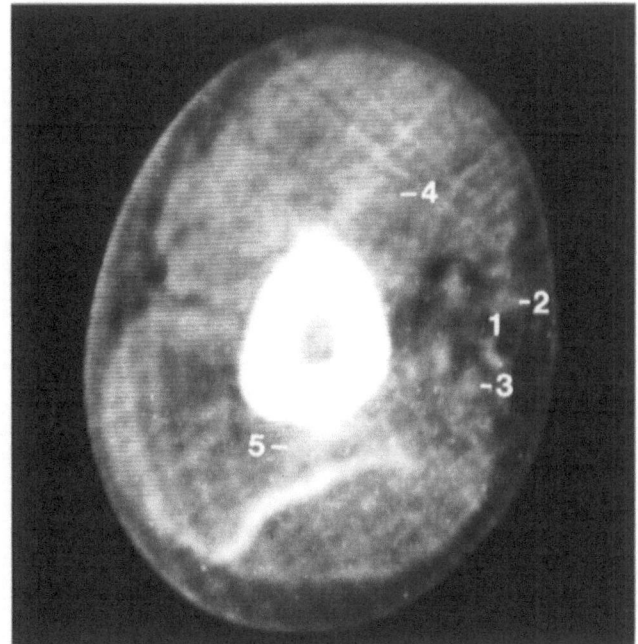

Distal diaphysis

1 Brachial vessels
2 Median nerve
3 Ulnar nerve
4 Musculocutaneous nerve
5 Radial nerve

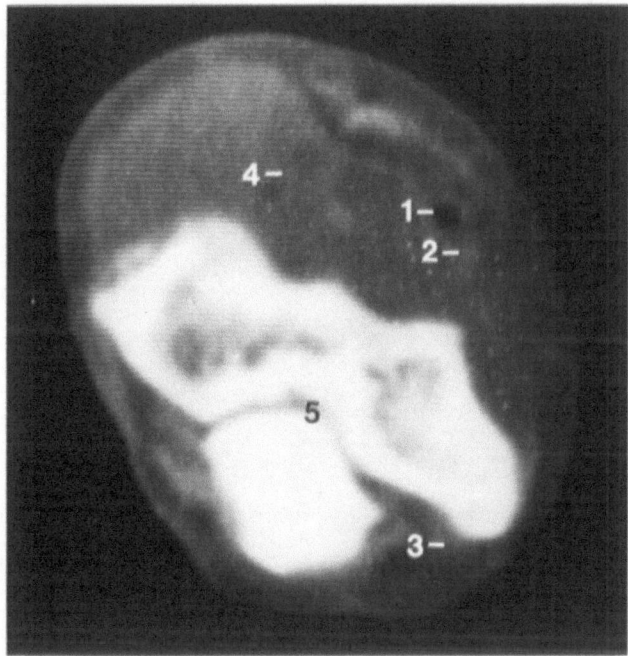

Distal epiphysis

1 Brachial vessels
2 Median nerve
3 Ulnar nerve
4 Radial nerve
5 Elbow joint

B. Cross-sections of the Forearm

Levels of the Cross-sections

The eight cross-sections of the forearm are arranged in four groups corresponding to areas commonly used for external fixation.

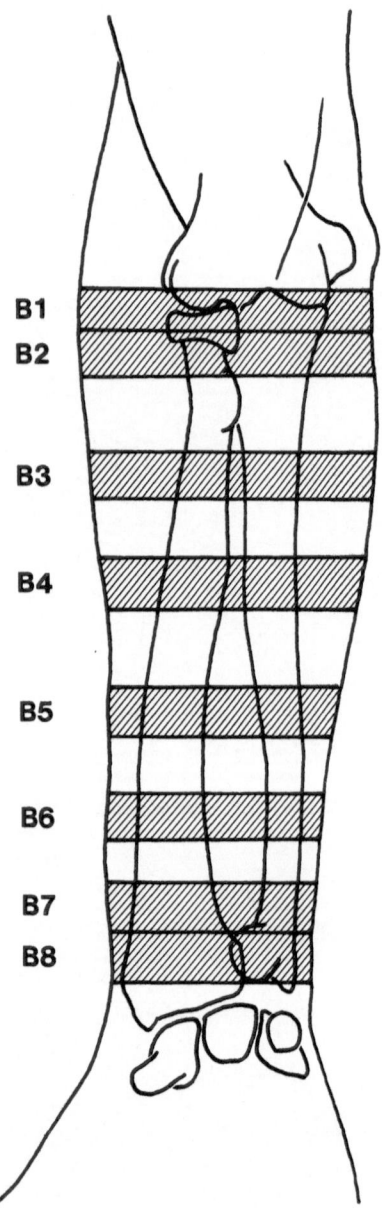

Proximal Epiphysis and Metaphysis (B 1, B 2)

Section B 1 shows the radial head and the coronoid process, while Section B 2 has been taken at the level of the radial neck.

Proximal Diaphysis (B 3, B 4)

The two sections, separated by 1.5 cm, have been taken at the level of the junction of the proximal and middle thirds of the radial diaphysis.

Distal Diaphysis (B 5, B 6)

The two sections, separated by 1.5 cm, have been taken at the level of the junction of the middle and distal thirds of radial diaphysis.

Distal Epiphysis and Metaphysis (B 7, B 8)

The two sections have been taken at the distal end of the forearm. Section B 7 has been taken close to the distal radio-ulnar joint, whereas Section B 8 passes through the radio-carpal joint.

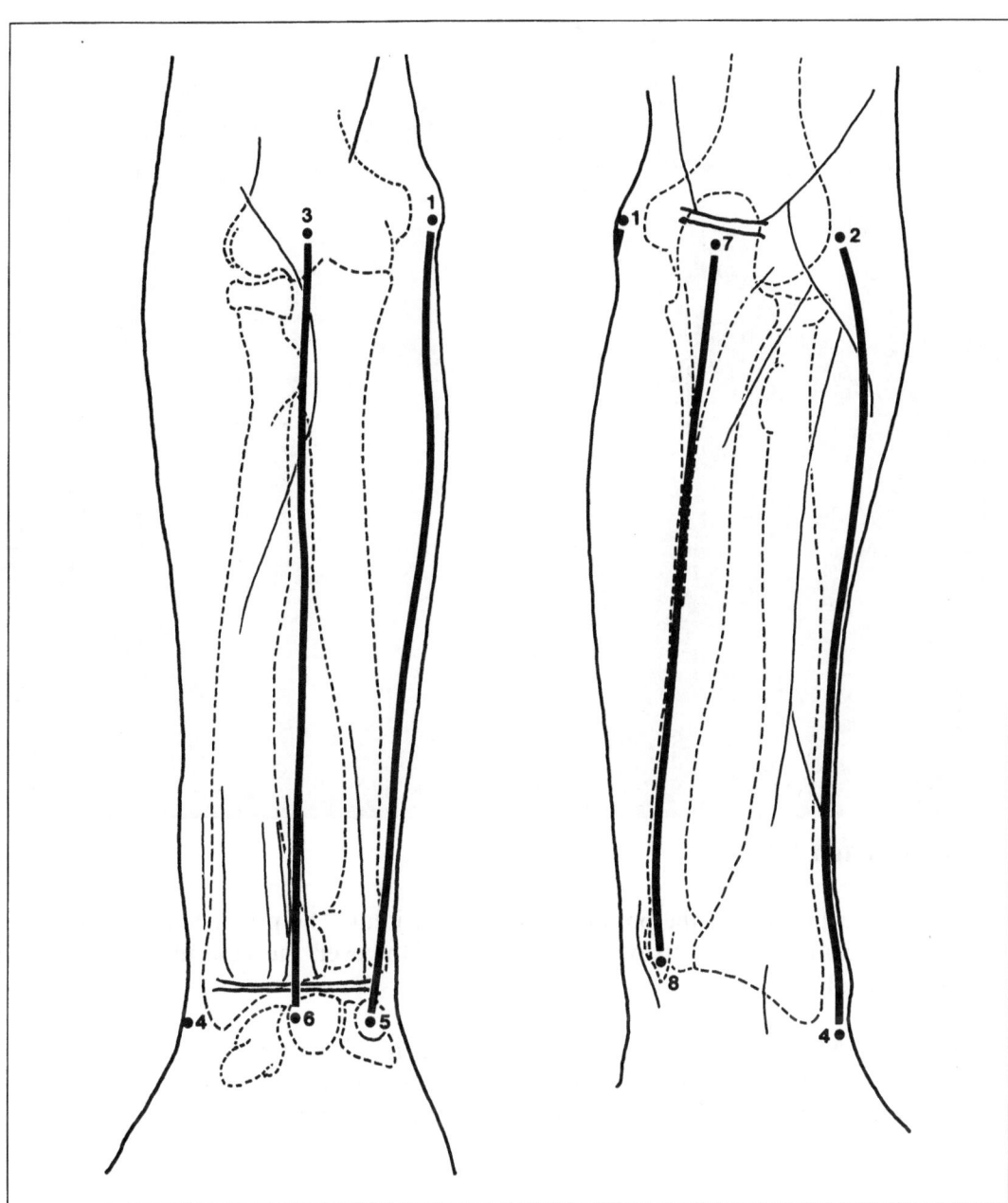

Landmarks

1 Medial epicondyle
2 Lateral epicondyle
3 Midpoint of line 1–2
4 Styloid process of radius
5 Pisiform bone
6 Midpoint of line 4–5
7 Olecranon
8 Styloid process of ulna

Reference lines

Ventral line:	line 3–6
Lateral line:	line 2–4
Medial line:	line 1–5
Dorsal border of ulna:	line 7–8

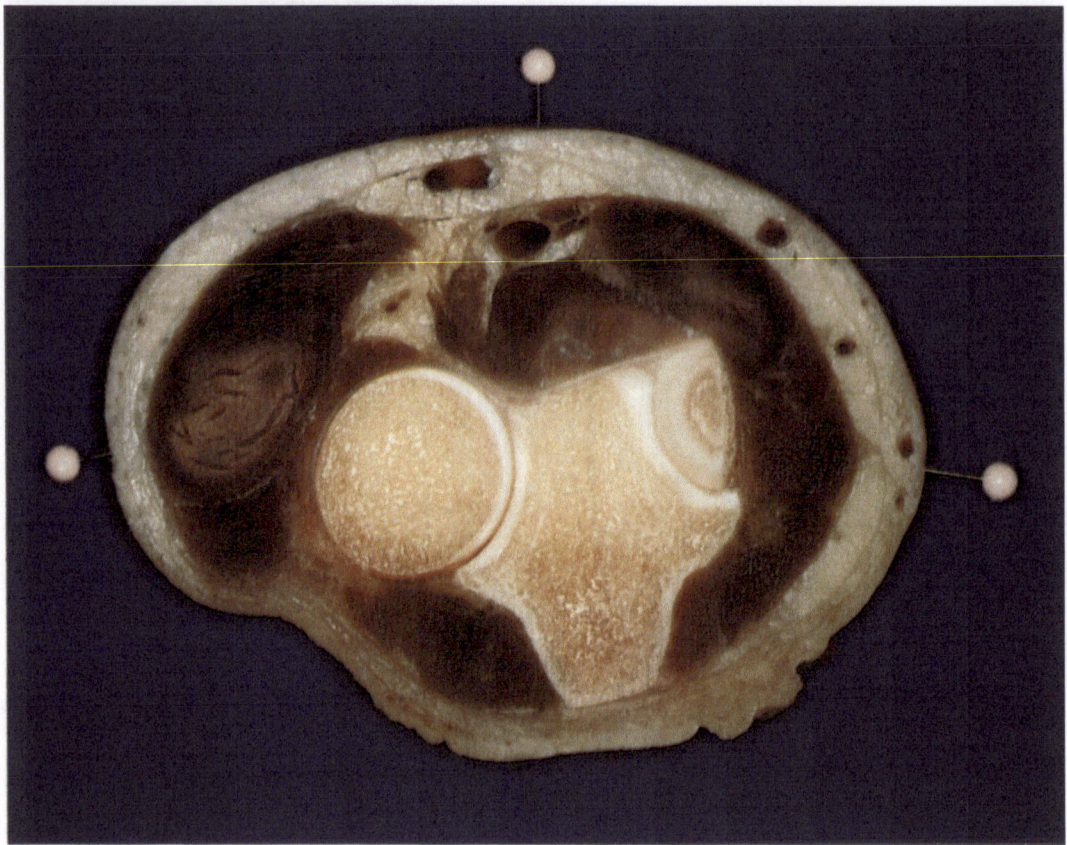

Cross-section B 1

Bone. The section shows the proximal radio-ulnar joint. The presence of this joint precludes the combined transfixion of both bones. However, the ulna may be transfixed through the olecranon.

Vessels and Nerves. The brachial vessels and the median and radial nerves lie close together, separated by the bicipital tendon. The ulnar nerve is situated at some distance from these structures, against the medial aspect of the olecranon.

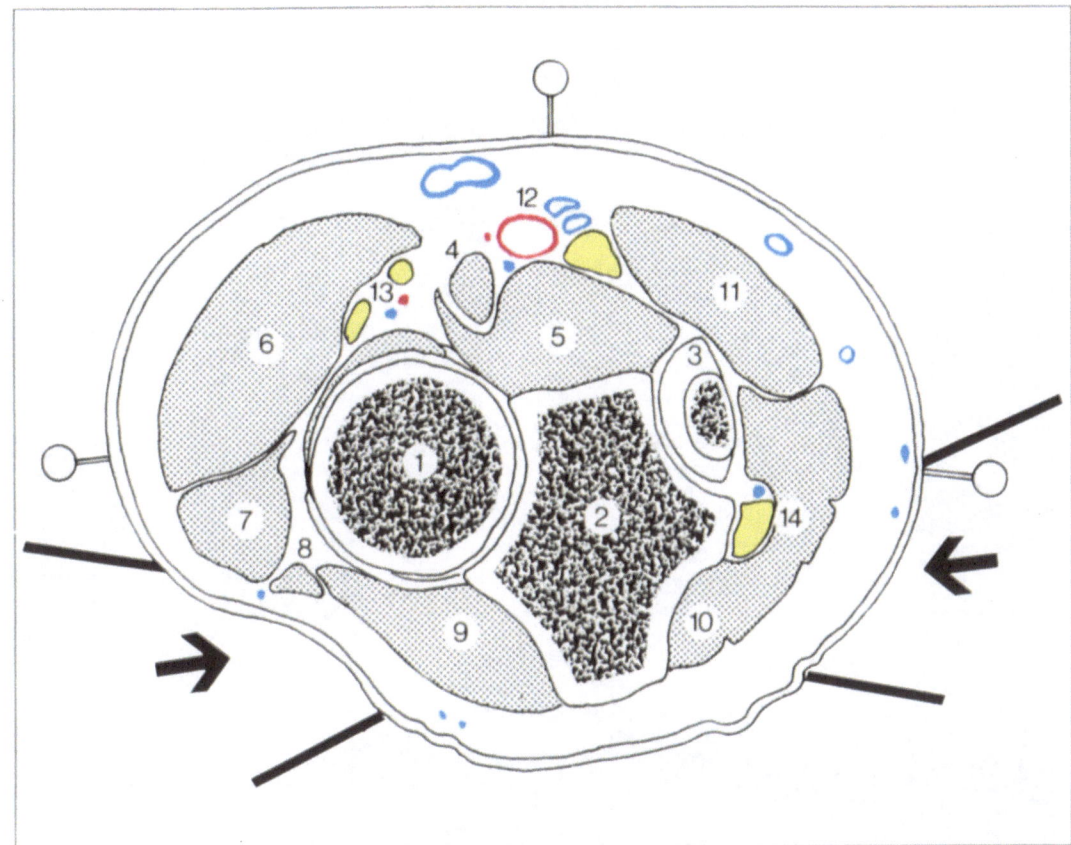

1 Radial head	7 Extensor carpi radialis longus	11 Pronator teres + flexor digitorum
2 Ulna	8 Extensor carpi radialis brevis +	superficialis + flexor carpi radialis
3 Humerus (trochlea)	extensor digitorum communis +	12 Brachial artery and vein +
4 Biceps brachii	extensor carpi ulnaris	median nerve
5 Brachialis	9 Anconeus	13 Radial nerve
6 Brachioradialis	10 Flexor carpi ulnaris	14 Ulnar nerve

Safe Areas. These are narrow and are situated medially and laterally in relation to the olecranon.

Cutaneous Zones Related to the Safe Areas. These comprise two narrow zones lying dorso-laterally and dorso-medially in relation to the reference lines.

- *The dorso-lateral zone* corresponds to the middle third of the area between the dorsal ulnar border and the lateral reference line.

- *The dorso-medial zone* corresponds to the ventral half of the area between the dorsal ulnar border and the medial reference line.

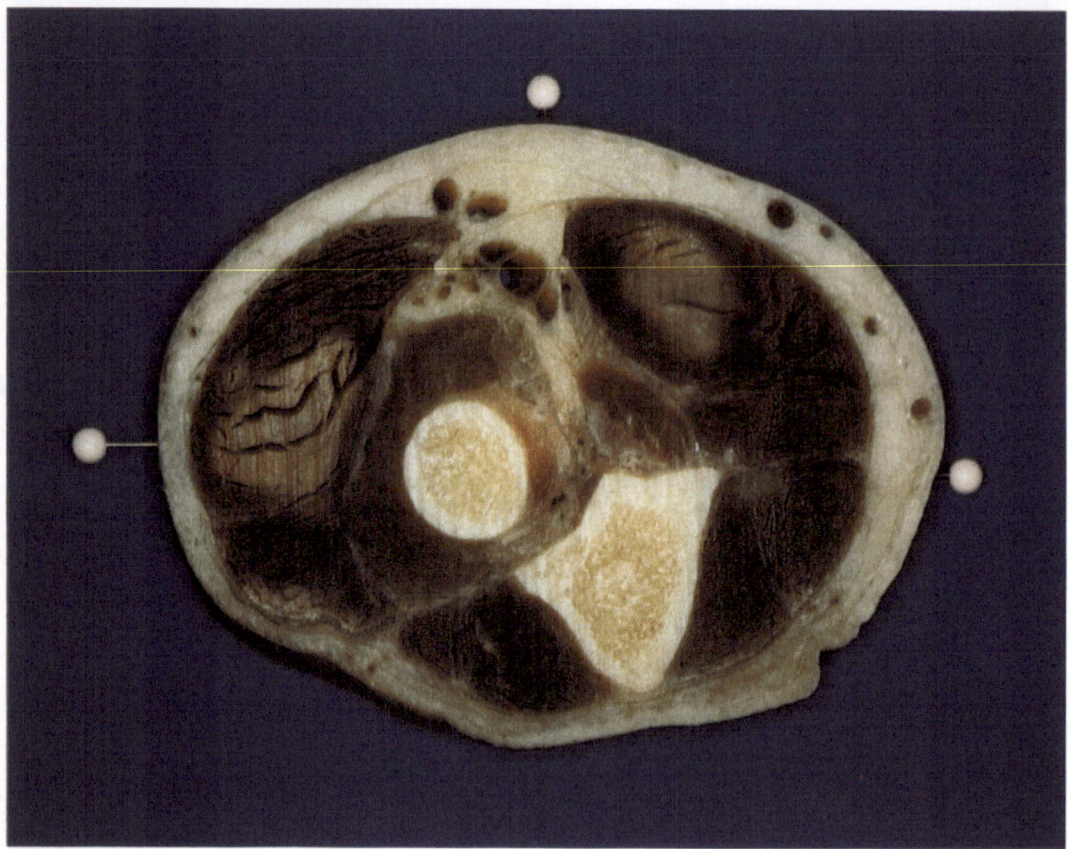

Cross-section B 2

Bone. The section has been taken at the level of the radial neck. Combined external fixation of both bones is not possible at this level owing to the intra-articular position of this section.

Vessels and Nerves. The brachial artery has not yet divided, but lies medially alongside the median nerve. The deep branch of the radial nerve lies deep to the bodies of the brachioradialis and extensor carpi radialis longus muscles between the two heads of the supinator muscle. The superficial radial nerve is covered by the medial border of the body of the brachioradialis muscle. The ulnar nerve passes alongside the ventral border of the ulnar metaphysis.

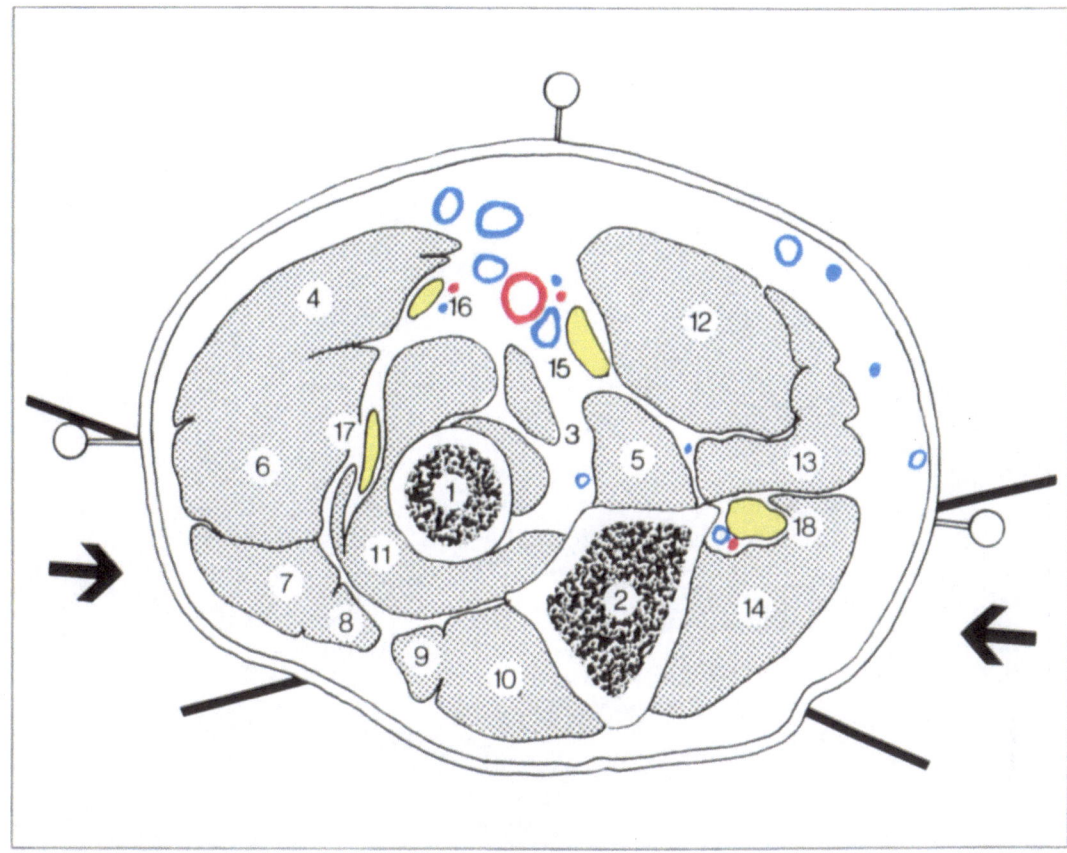

1 Radius
2 Ulna
3 Biceps brachii
4 Brachioradialis
5 Brachialis
6 Extensor carpi radialis longus
7 Extensor carpi radialis brevis
8 Extensor digitorum communis
9 Extensor digiti minimi
10 Extensor carpi ulnaris

11 Supinator
12 Pronator teres
13 Flexor digitorum superficialis + flexor carpi ulnaris
   + flexor carpi radialis
14 Flexor carpi ulnaris
15 Median nerve + brachial vessels
16 Radial nerve (superficial branch)
17 Radial nerve (deep branch)
18 Ulnar nerve

Safe Areas. These are narrow and lie medially and laterally in relation to the ulna.

Cutaneous Zones Related to the Safe Areas. These form two zones lying dorso-laterally and dorso-medially in relation to the reference lines.

- *The dorso-lateral zone* corresponds to the middle third of the area between the dorsal ulnar border and the lateral reference line.

- *The dorso-medial zone* corresponds to the ventral half of the area between the dorsal ulnar border and the medial reference line.

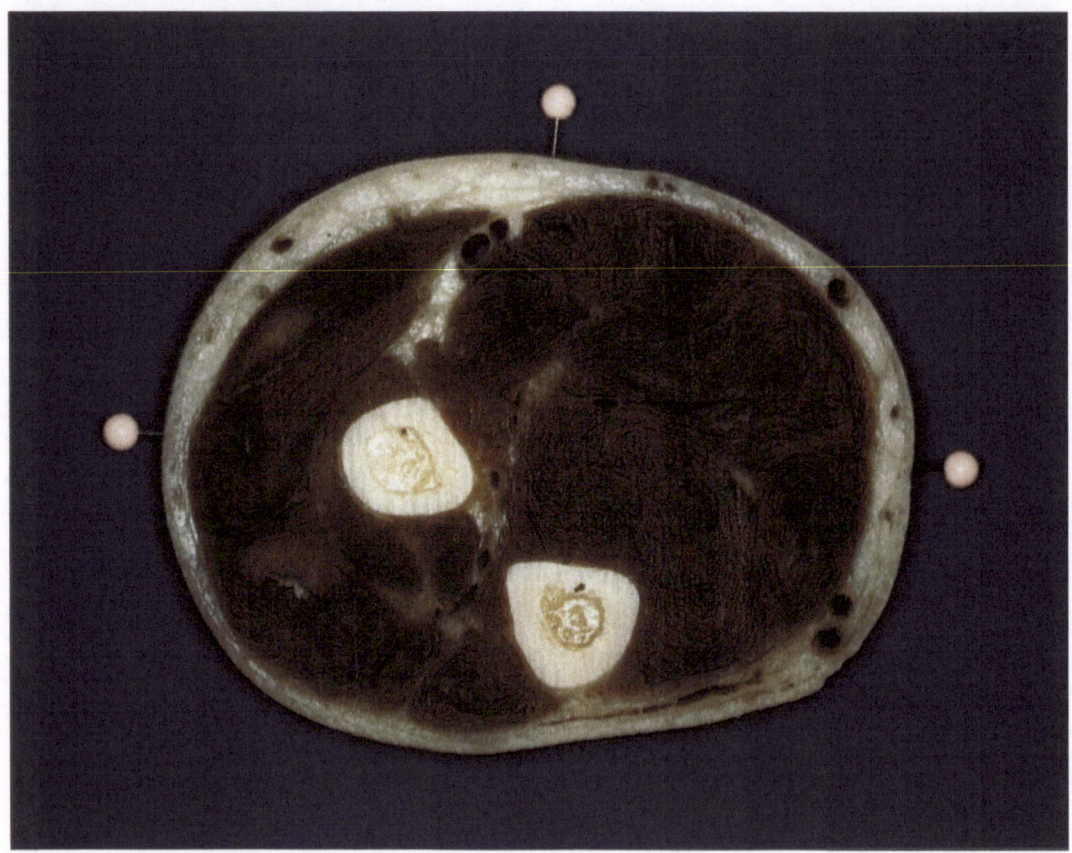

Cross-section B 3

Bone. The section has been taken proximal to the junction of the proximal and middle thirds of the radial diaphysis. The radius lies deep within the forearm, whereas the dorsal border of the ulna is located at the mid-dorsal aspect of the forearm.

Vessels and Nerves. The radial and ulnar arteries and the ulnar nerve lie ventrally in a plane parallel to the two bones.

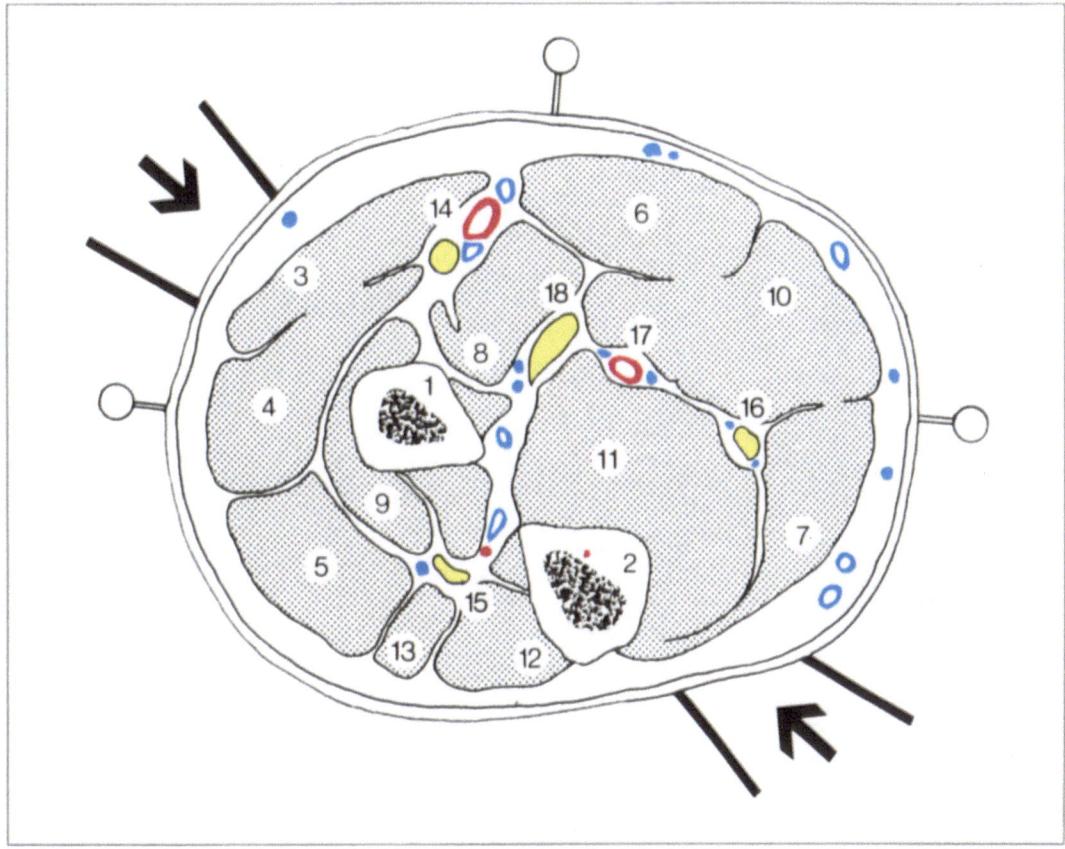

1 Radius
2 Ulna
3 Brachioradialis
4 Extensor carpi radialis longus
5 Extensor carpi radialis brevis
6 Flexor carpi radialis
7 Flexor carpi ulnaris
8 Pronator teres
9 Supinator
10 Flexor digitorum superficialis
11 Flexor digitorum profundus
12 Extensor carpi ulnaris
13 Extensor digitorum communis + extensor digiti minimi
14 Radial vessels + radial nerve (superficial branch)
15 Radial nerve (deep branch)
16 Ulnar nerve
17 Ulnar vessels
18 Median nerve

Combined External Fixation of Both Bones of the Forearm

Safe Areas. These are narrow and lie ventro-laterally and dorso-medially in relation to the radius and ulna.

Cutaneous Zones Related to the Safe Areas. These comprise two zones lying ventro-laterally and dorso-medially in relation to the reference lines.

- *The ventro-lateral zone* corresponds to the dorsal half of the body of the brachioradialis muscle.

- *The dorso-medial zone* corresponds to the dorsal half of the flexor carpi ulnaris muscle facing the medial aspect of the ulna.

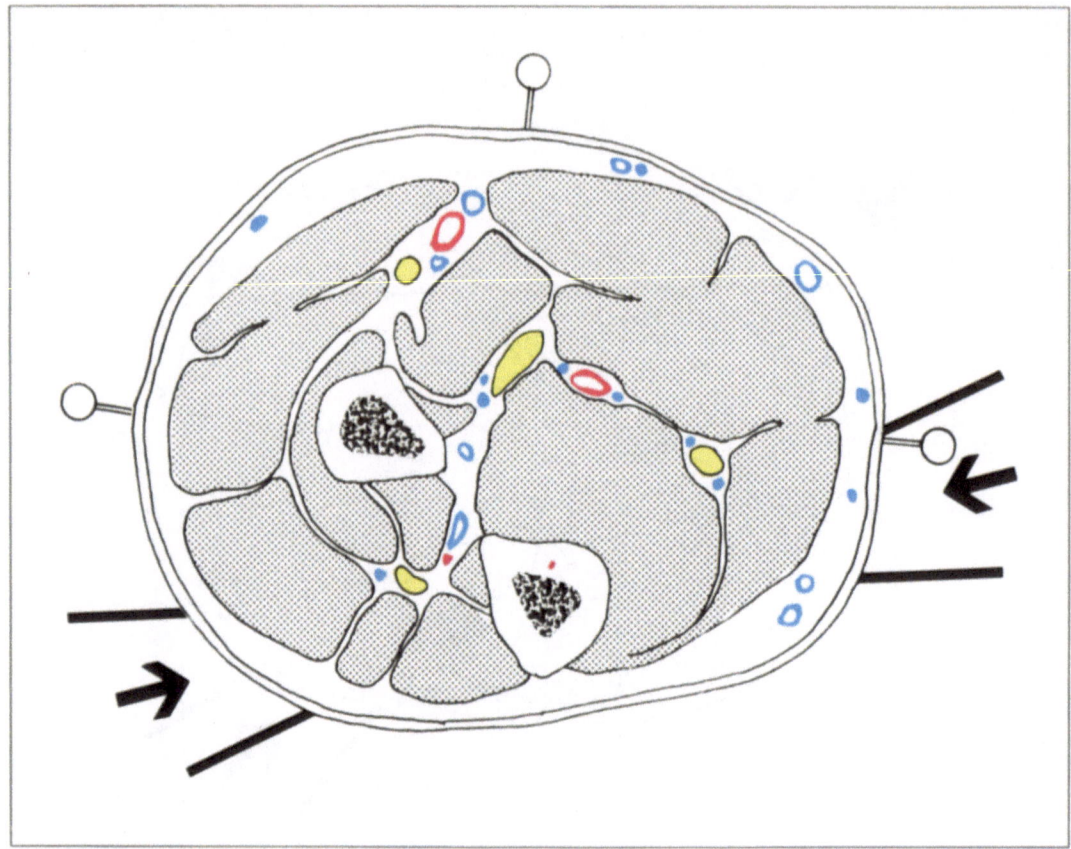

Cross-section B 3

Separate External Fixation of the Ulna

Safe Areas. These are very narrow and lie medially and laterally in relation to the ulna.

Cutaneous Zones Related to the Safe Areas. These comprise two zones lying dorso-laterally and dorso-medially in relation to the reference lines.

- *The dorso-lateral zone* corresponds to the middle third of the area between the dorsal border of the ulna and the lateral reference line. This corresponds to the body of the extensor carpi radialis brevis muscle.

- *The dorso-medial zone* corresponds to the ventral half of the area between the dorsal border of the ulna and the medial reference line. This corresponds to the ventral half of the body of the flexor carpi ulnaris muscle.

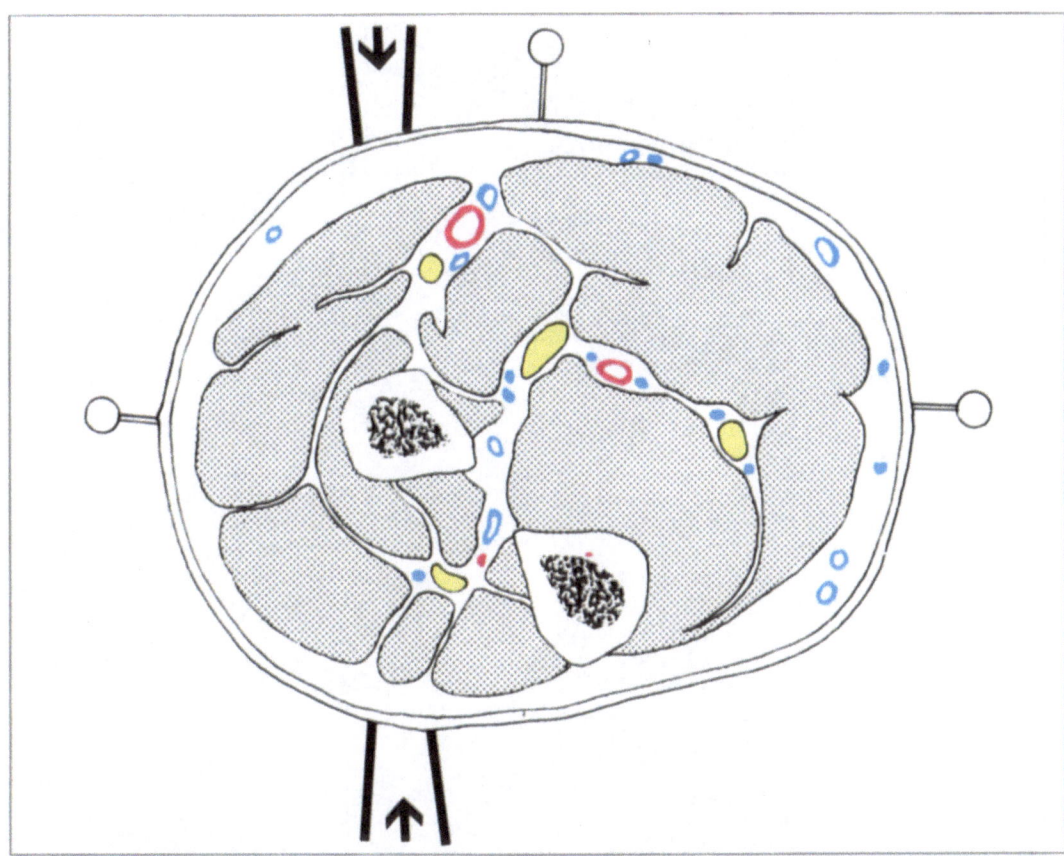

Cross-section B 3

Separate External Fixation of the Radius

Safe Areas. These are much reduced and lie ventrally and dorsally in relation to the radius.

Cutaneous Zones Related to the Safe Areas. These comprise two very narrow zones lying lateral to the ventral reference line and to the dorsal border of the ulna.
These two zones are reduced to a small area centred on the sagittal axis of the radial diaphysis.

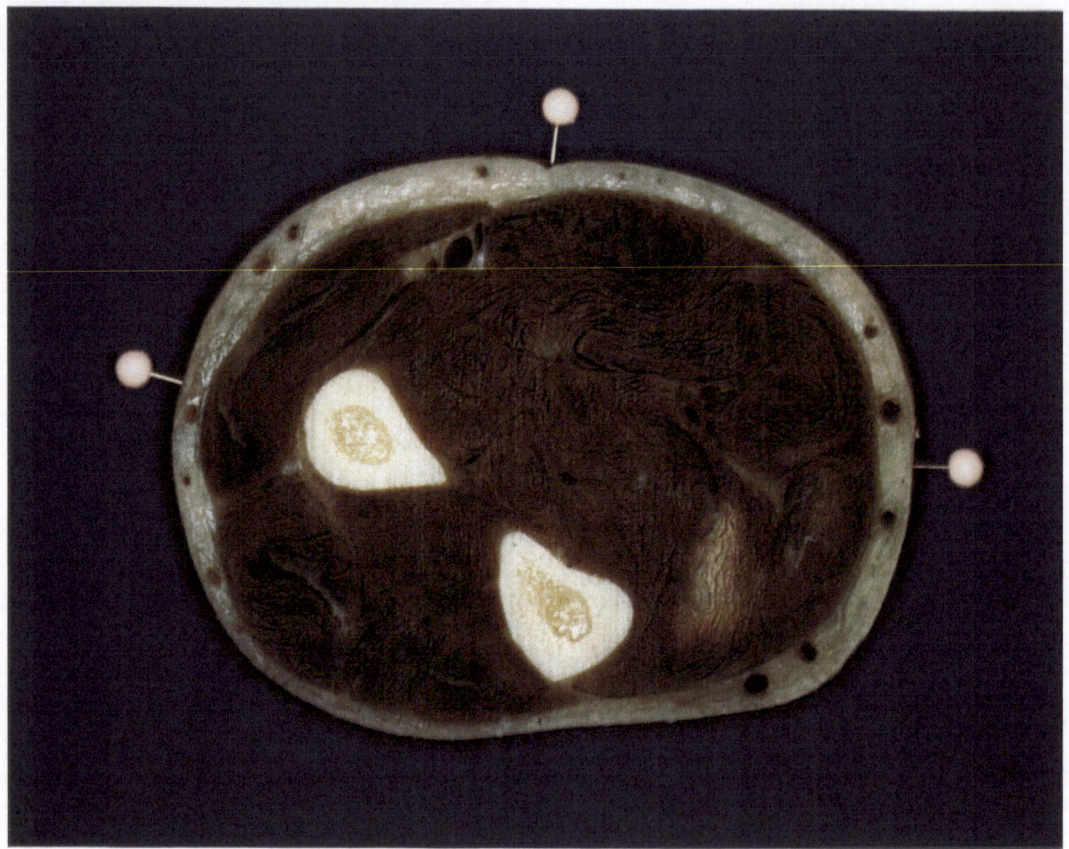

Cross-section B 4

Bone. The section has been taken distal to the junction of the proximal and middle thirds of the radial diaphysis. The radius lies deep within the forearm, and the dorsal border of the ulna lies in the middle of the dorsal aspect of the forearm.

Vessels and Nerves. The radial and ulnar vessels, the branches of the superficial radial nerve, and the median and ulnar nerves all lie in a plane parallel to the two bones. The deep branch of the radial nerve divides into muscular branches supplying the dorsal muscles of the forearm.

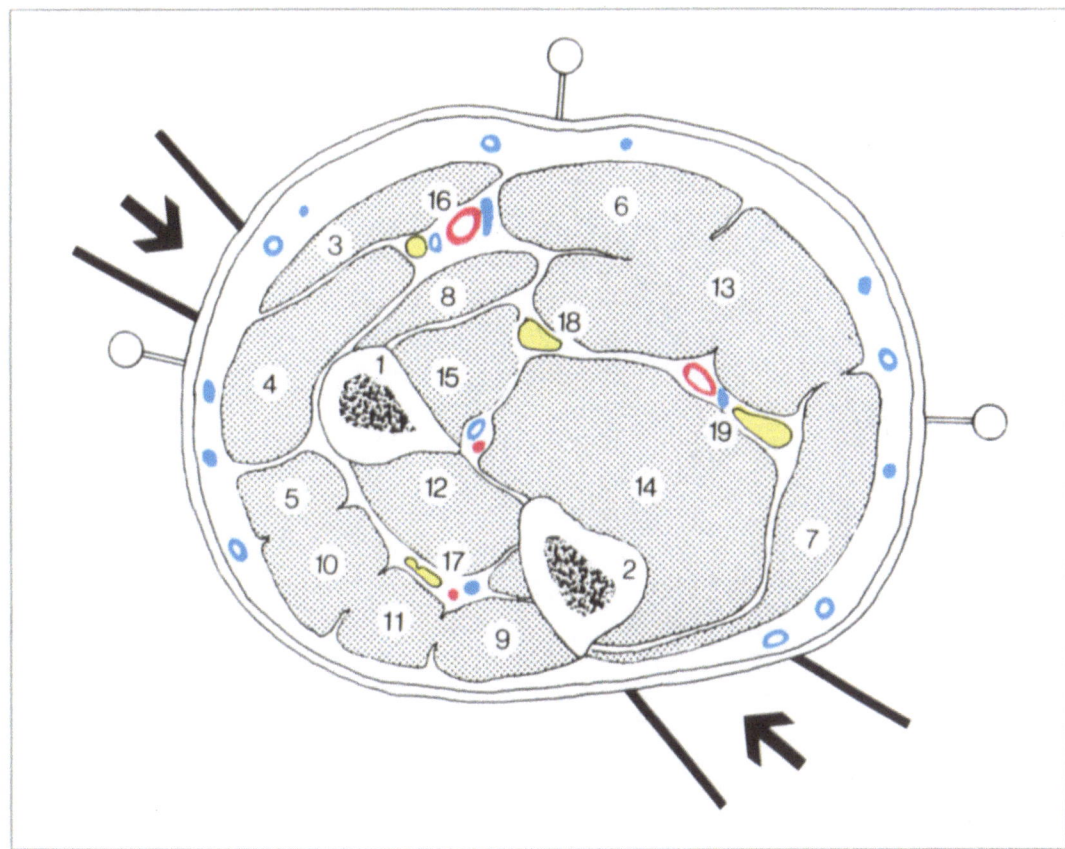

1 Radius	11 Extensor digiti minimi
2 Ulna	12 Abductor pollicis longus
3 Brachioradialis	13 Flexor digitorum superficialis
4 Extensor carpi radialis longus	14 Flexor digitorum profundus
5 Extensor carpi radialis brevis	15 Flexor pollicis longus
6 Flexor carpi radialis	16 Radial vessels + radial nerve (superficial branch)
7 Flexor carpi ulnaris	17 Radial nerve (muscular branches)
8 Pronator teres	18 Median nerve
9 Extensor carpi ulnaris	19 Ulnar vessels + ulnar nerve
10 Extensor digitorum communis	

Combined External Fixation of Both Bones of the Forearm

Safe Areas. These are very narrow owing to the slenderness of the radial and ulnar diaphyses and lie ventro-laterally and dorso-medially in relation to the two bones.

Cutaneous Zones Related to the Safe Areas. These comprise two narrow zones lying ventrolaterally and dorso-medially in relation to the reference lines.

- *The ventro-lateral zone* is very narrow and lies ventral and adjacent to the ventral reference line. The zone corresponds to the lateral aspect of the radius, but is more superficial.

- *The narrow dorso-medial zone* lies in the dorsal third of the area between the dorsal ulnar border and the medial reference line. The zone corresponds to the dorsal third of the body of the flexor carpi ulnaris muscle facing the medial aspect of the ulna.

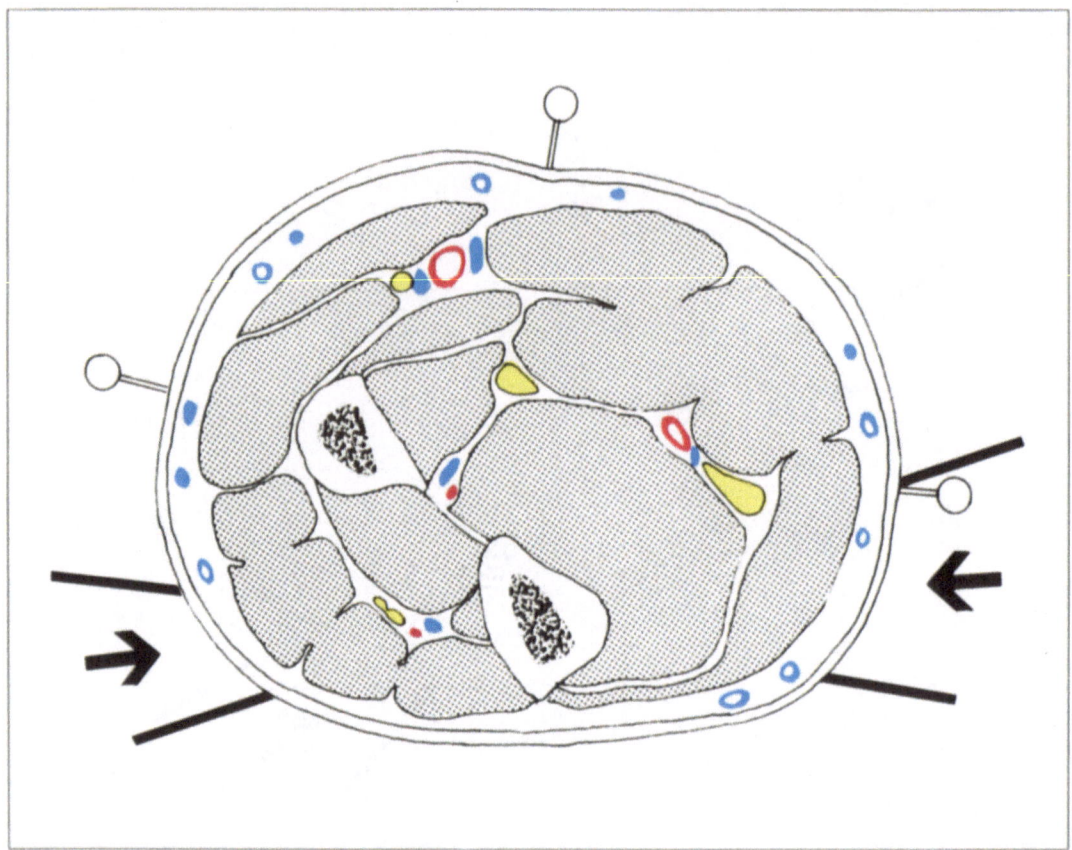

Cross-section B 4

Separate External Fixation of the Ulna

Safe Areas. These are narrow owing to the disposition of the nerves and arteries; they lie medially and dorso-laterally in relation to the ulna.

Cutaneous Zones Related to the Safe Areas. These comprise two narrow zones lying dorso-laterally and dorso-medially in relation to the reference lines.

- *The dorso-lateral zone* consists of a quarter of the area lying between the lateral reference line and the dorsal border of the ulna. This zone lies in the middle of this area and faces the body of the extensor digitorum communis muscle.

- *The dorso-medial zone* consists of the ventral half of the area lying between the medial reference line and the dorsal border of the ulna. It overlies the ventral half of the flexor carpi ulnaris muscle, stopping short of the ventral border of this muscle.

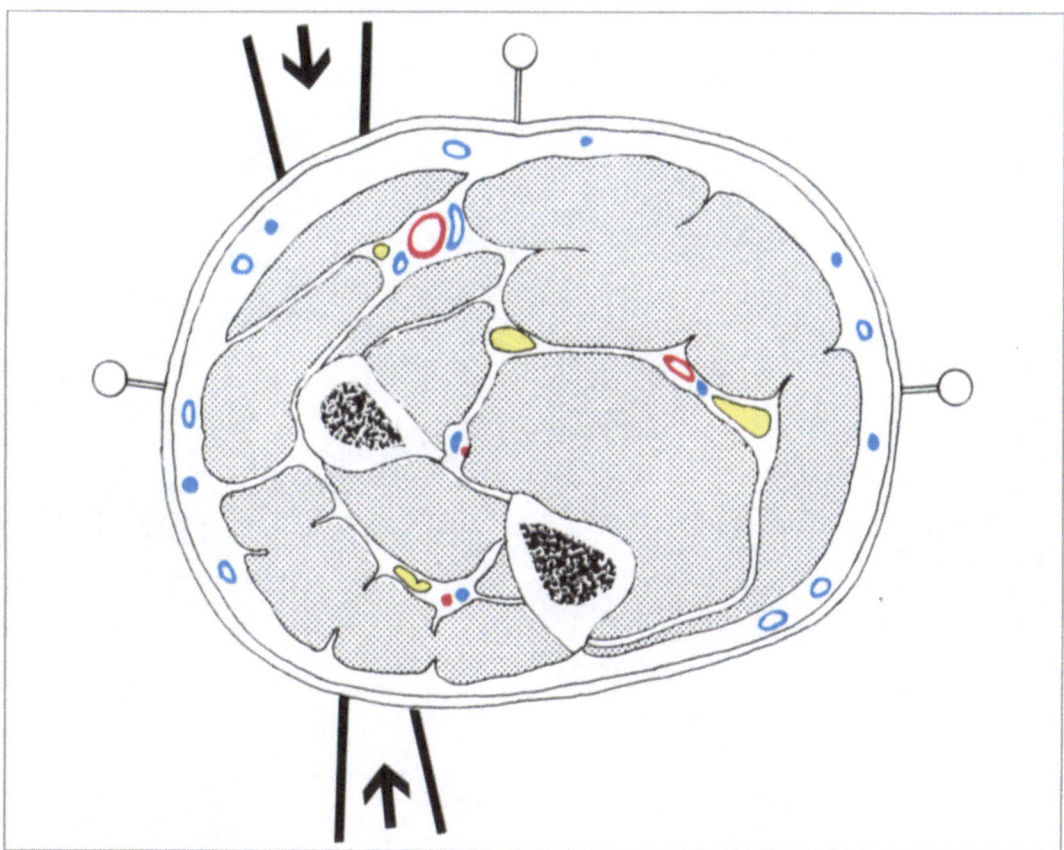

Cross-section B 4

Separate External Fixation of the Radius

Safe Areas. These are much reduced and lie ventrally and dorsally in relation to the radius.

Cutaneous Zones Related to the Safe Areas. These comprise two very narrow zones lying lateral to the ventral reference line and to the dorsal border of the ulna.
These two zones are reduced to a small area centred on the sagittal axis of the radial diaphysis.

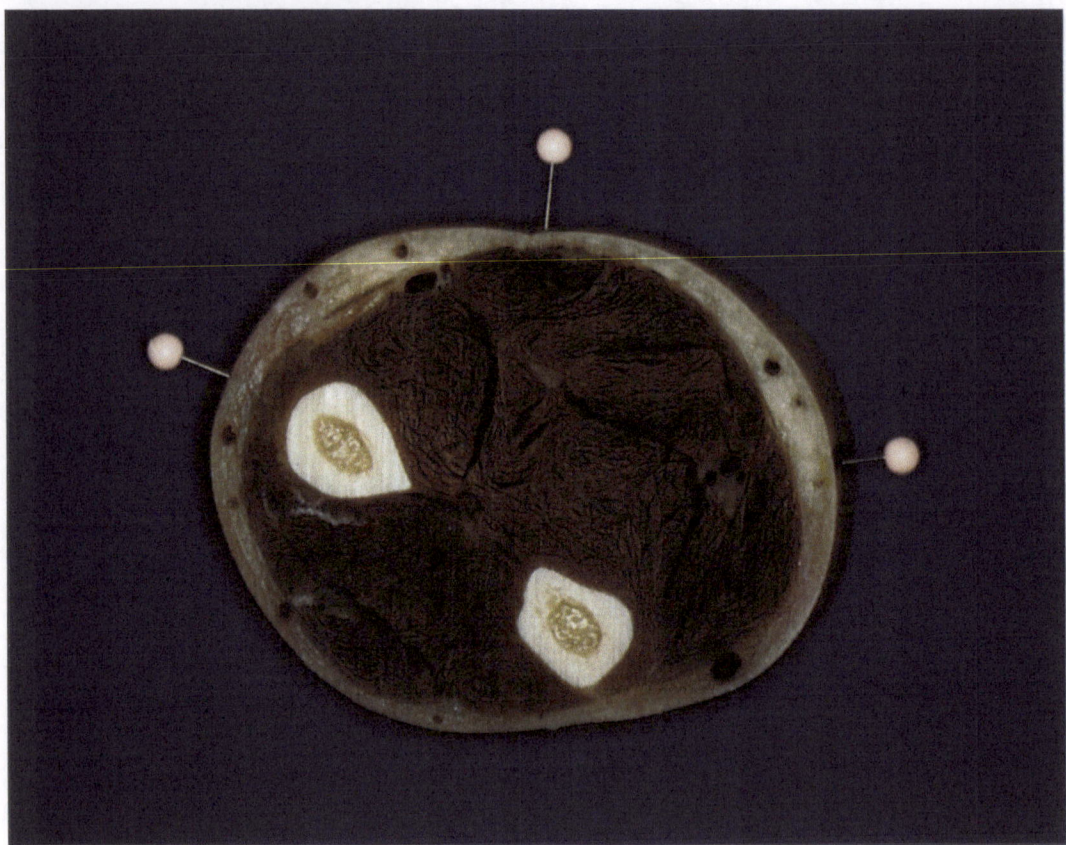

Cross-section B 5

Bone. The section has been taken immediately proximal to the junction of the middle and distal thirds of the radius. The radius is now superficial and the ulna more medial.

Vessels and Nerves. The radial and ulnar vessels and the median and ulnar nerves all lie in a ventral plane parallel to the bones. The deep branch of the radial nerve ramifies towards the dorsal forearm muscles.

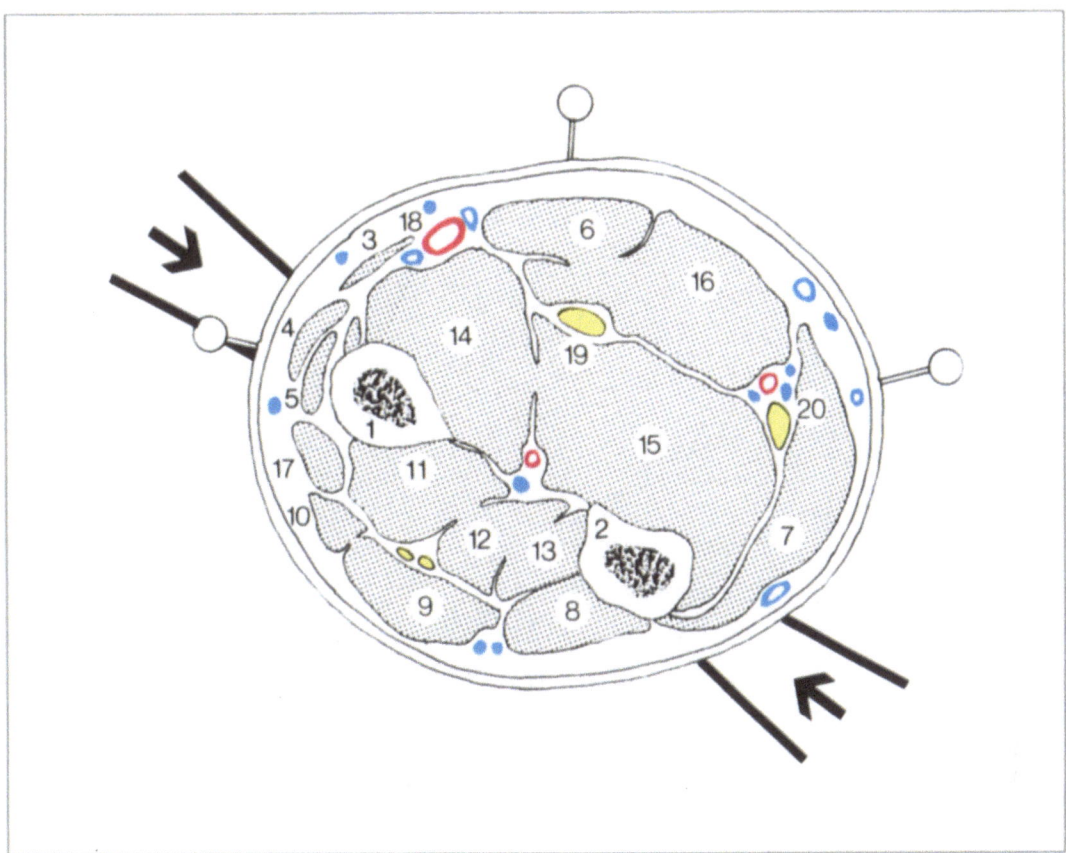

1 Radius	8 Extensor carpi ulnaris	14 Flexor pollicis longus
2 Ulna	9 Extensor digitorum	15 Flexor digitorum profundus
3 Brachioradialis	communis	16 Flexor digitorum superficialis
4 Extensor carpi radialis longus	10 Extensor digiti minimi	17 Abductor pollicis
5 Extensor carpi radialis brevis	11 Extensor pollicis longus	18 Radial vessels
6 Flexor carpi radialis	12 Extensor pollicis brevis	19 Median nerve
7 Flexor carpi ulnaris	13 Extensor indicis	20 Ulnar vessels + ulnar nerve

Combined External Fixation of Both Bones of the Forearm

Safe Areas. These are narrow and lie ventro-laterally and dorso-medially in relation to the radius and ulna.

Cutaneous Zones Related to the Safe Areas. These comprise two zones lying ventro-laterally and dorso-medially in relation to the reference lines.

- *The ventro-lateral zone* lies between the lateral aspect of the radius, which lies superficially beneath the tendons of the extensor carpi radialis brevis muscle, and the lateral reference line.

- *The dorso-lateral zone* lies adjacent to the dorsal ulnar border and the dorsal part of the flexor carpi ulnaris muscle and corresponds to the medial aspect of the ulna.

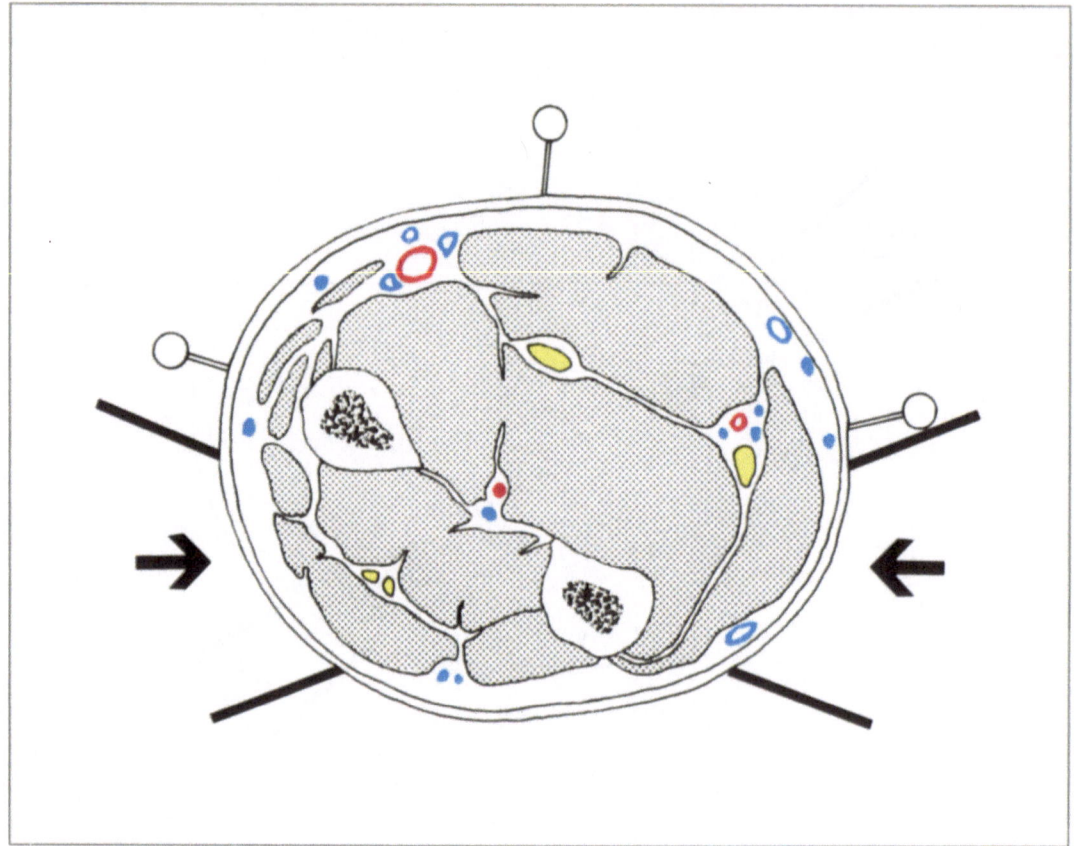

Cross-section B 5

Separate External Fixation of the Ulna

Safe Areas. These have become more extensive owing to the ramification of the deep branch of the radial nerve and the ventral migration of the ulnar nerve. The areas lie medially and laterally in relation to the ulna.

Cutaneous Zones Related to the Safe Areas. These comprise two zones lying dorso-medially and dorso-laterally in relation to the reference lines.

- *The dorso-lateral zone* consists of half of the area between the lateral reference line and the dorsal border of the ulna. Its lateral boundary lies slightly dorsal to the lateral reference line to avoid the radius.

- *The dorso-medial zone* consists of the ventral half of the area between the dorsal ulnar border and the medial reference line, which it does not quite reach.

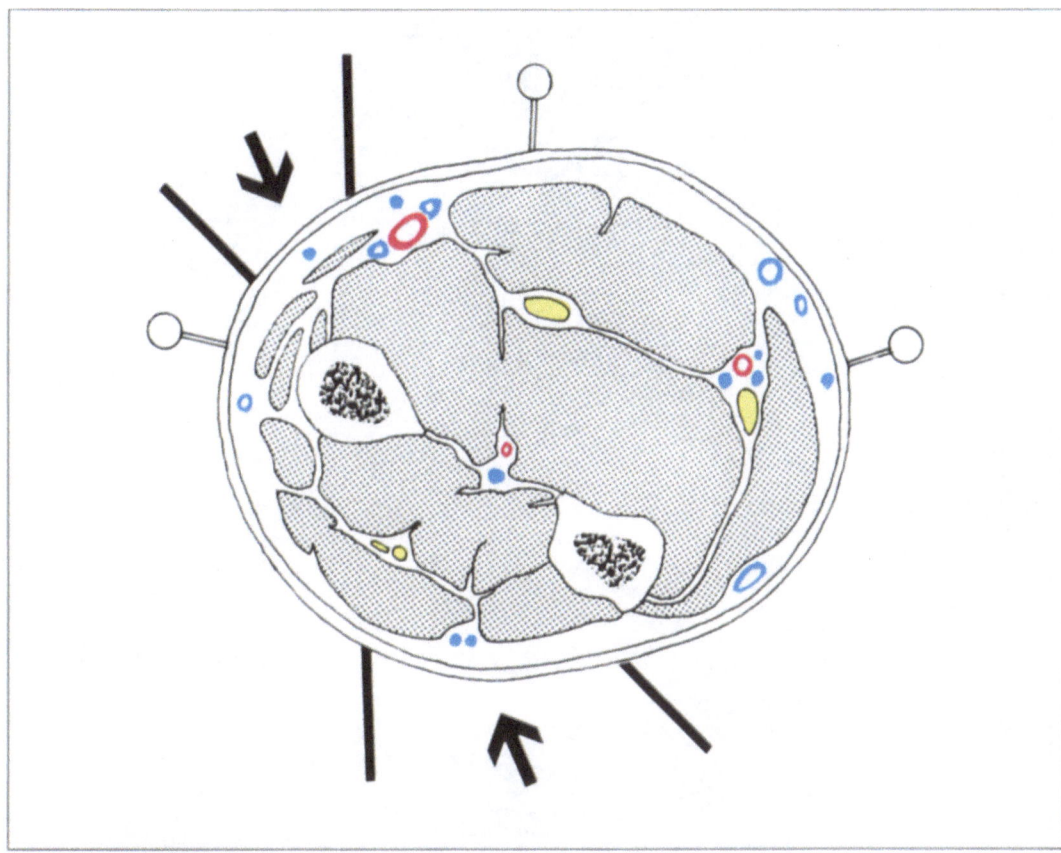

Cross-section B 5

Separate External Fixation of the Radius

Safe Areas. These have become much larger owing to the lateral displacement of the radius and the ramification of the deep branch of the radial nerve. The areas lie ventrally and dorsally in relation to the radial diaphysis.

Cutaneous Zones Related to the Safe Areas. These comprise two zones lying ventro-laterally and dorso-laterally in relation to the reference lines.

- *The ventro-lateral zone* lies between the tendon of the brachioradialis muscle and the mid-lateral aspect of the radius.

- *The dorso-lateral zone* consists of the dorsal half of the area lying between the lateral reference line and the dorsal border of the ulna.

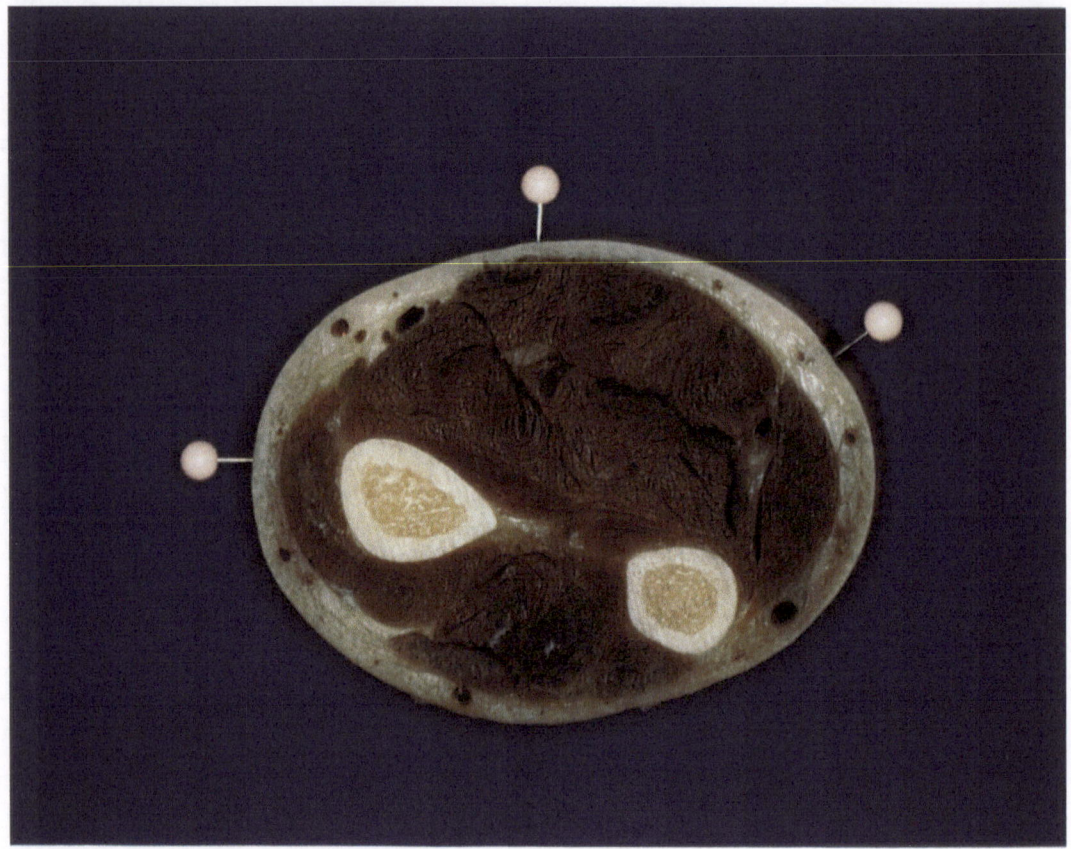

Cross-section B 6

Bone. The section has been taken immediately distal to the junction of the middle and distal thirds of the radius. The radius is now superficial and the ulna more medial.

Vessels and Nerves. The radial vessels are superficial and ventral, and the tendon of the brachioradialis muscle lies lateral to them. The ulnar vessels and nerve are covered by the flexor carpi ulnaris muscle.

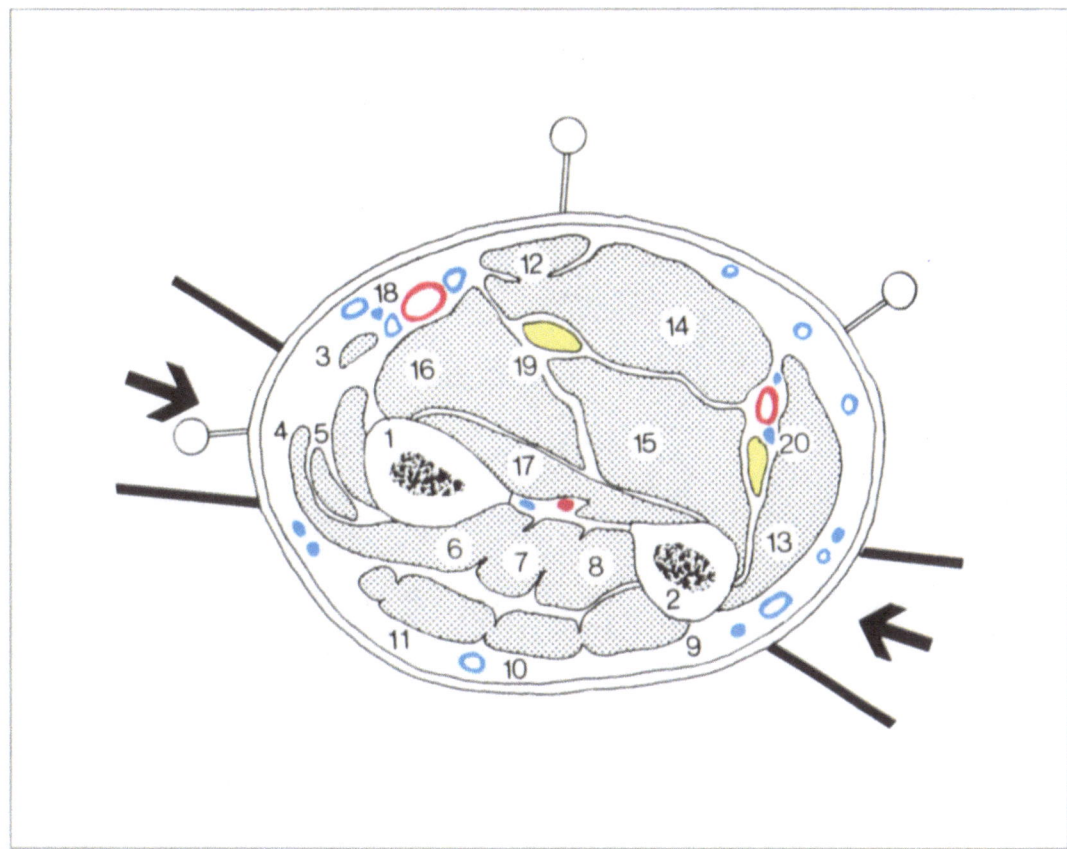

1 Radius	8 Extensor indicis	14 Flexor digitorum superficialis
2 Ulna	9 Extensor carpi ulnaris	15 Flexor digitorum profundus
3 Brachioradialis	10 Extensor digiti minimi	16 Flexor pollicis longus
4 Extensor carpi radialis longus	11 Extensor digitorum communis	17 Pronator quadratus
5 Extensor carpi radialis brevis	12 Flexor carpi radialis	18 Radial vessels
6 Abductor pollicis longus + extensor pollicis brevis	13 Flexor carpi ulnaris	19 Median nerve
7 Extensor pollicis longus		20 Ulnar vessels + ulnar nerve

Combined External Fixation of Both Bones of the Forearm

Safe Areas. These are ventro-lateral and dorso-medial in relation to the radius and ulna.

Cutaneous Zones Related to the Safe Areas. These comprise two zones ventro-lateral and dorso-medial in relation to the reference lines.

- *The ventro-lateral zone* lies on either side of the lateral reference line and consists of the area between the tendon of the brachioradialis muscle and that of the extensor carpi radialis brevis muscle. It is situated on the lateral aspect of the radius.

- *The dorso-medial zone* consists of the dorsal third of the area between the medial reference line and the dorsal border of the ulna, against which it abuts. It corresponds to the medial aspect of the ulna.

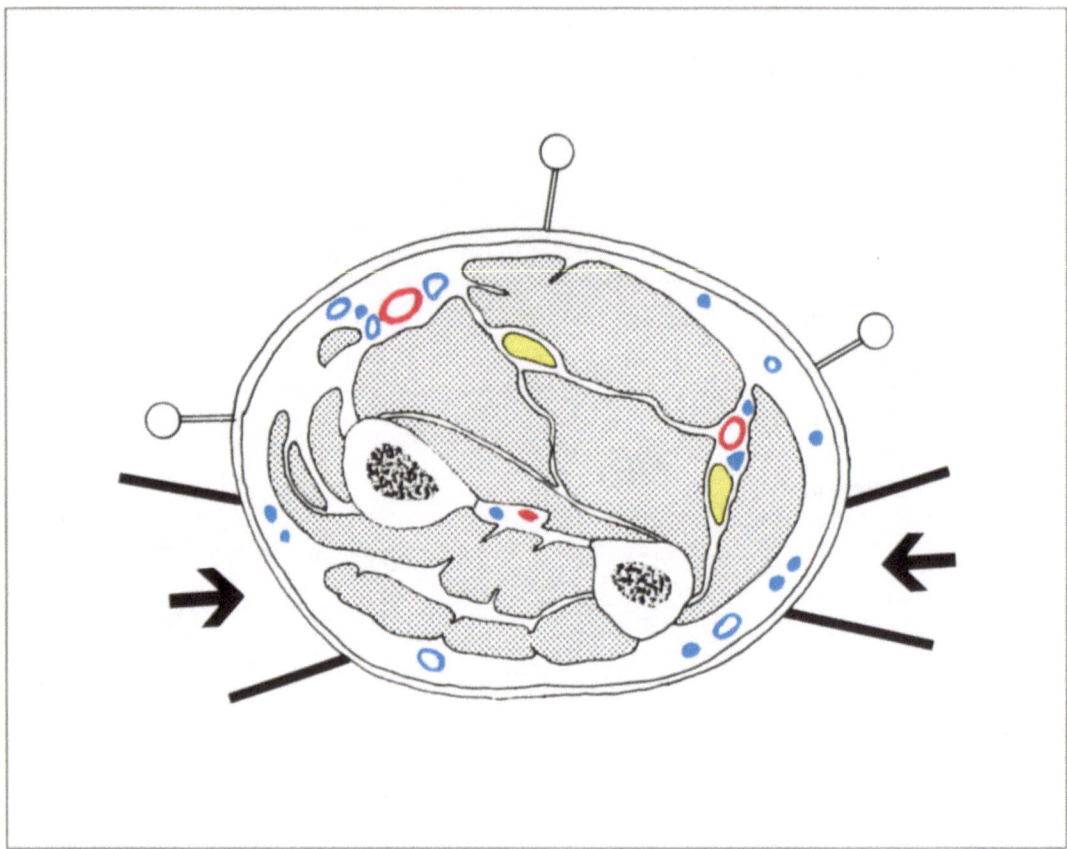

Cross-section B 6

Separate External Fixation of the Ulna

Safe Areas. These are slightly reduced because of the diminishing size of the forearm and are found medially and laterally in relation to the ulna.

Cutaneous Zones Related to the Safe Areas. These comprise two zones lying dorso-laterally and dorso-medially in relation to the reference lines.

- *The dorso-lateral zone* consists of the lateral half of the area between the dorsal border of the ulna and falls short of the lateral reference line.
- *The dorso-medial zone* consists of the middle third of the area between the medial reference line and the dorsal border of the ulna.

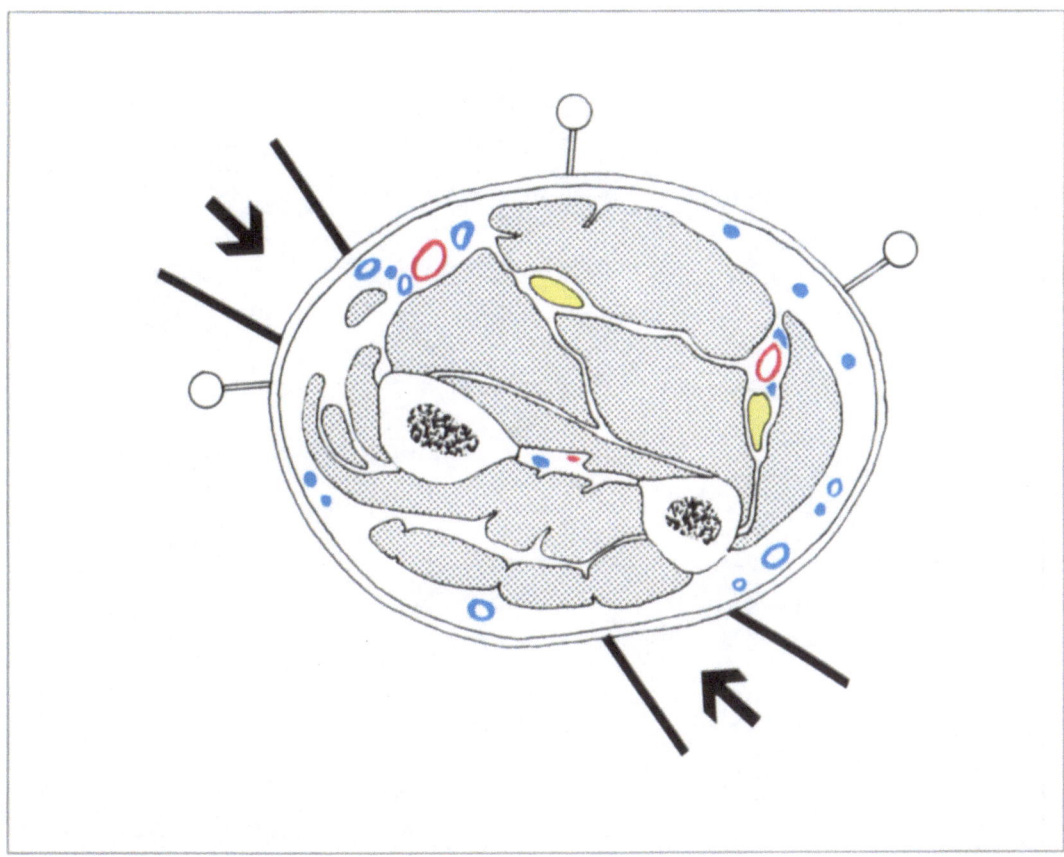

Cross-section B 6

Separate External Fixation of the Radius

Safe Areas. These are ventro-lateral and dorso-medial in relation to the radius.

Cutaneous Zones Related to the Safe Areas. These comprise two narrow zones lying ventro-laterally and dorsally in relation to the reference lines.

- *The ventro-lateral zone* lies between the tendon of the brachioradialis muscle and the lateral reference line.
- *The dorso-lateral zone* lies between the body of the extensor carpi ulnaris muscle and the dorsal border of the ulna.

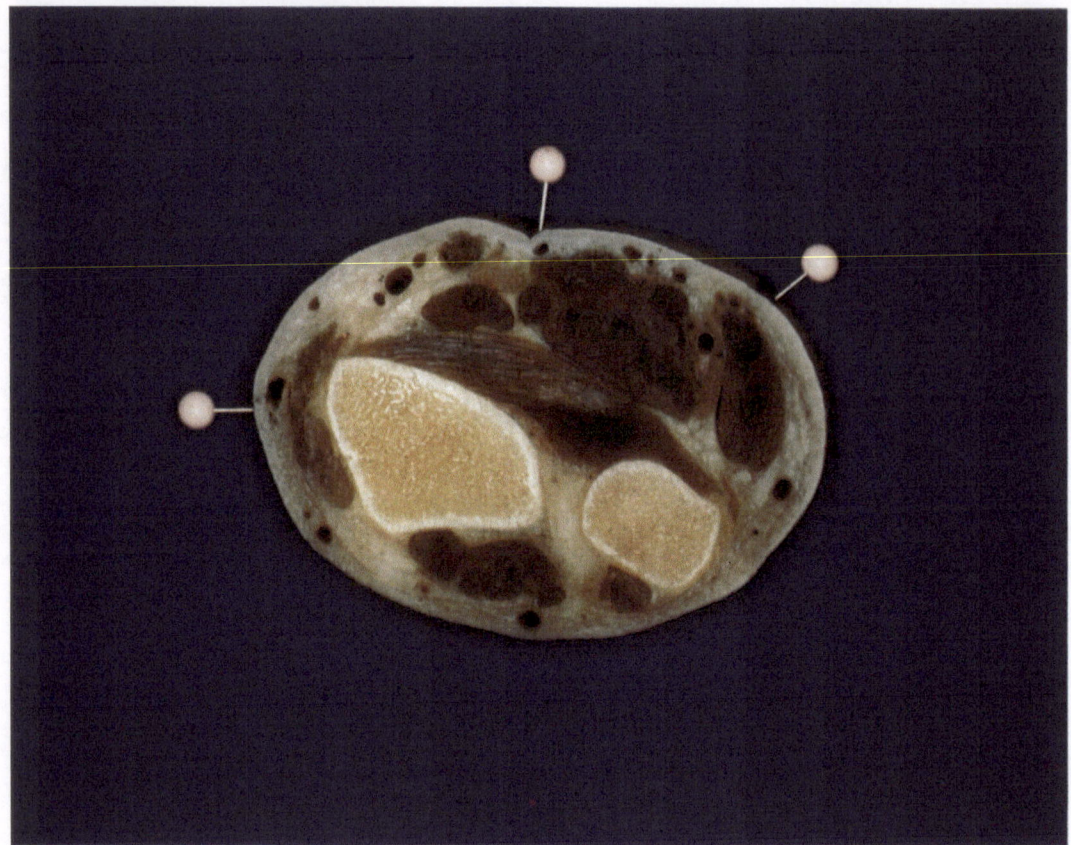

Cross-section B 7

Bones. The section has been taken through the distal radial metaphysis. The diameter of both bones has increased, but the proximity of the radio-ulnar joint does not permit combined external fixation at this level.

Vessels and Nerves. The radial vessels and the median nerve lie ventrally in relation to the radial metaphysis, and the ulnar vessels and nerve lie on the ventral aspect of the ulna. The digital flexor tendons lie between the two neurovascular bundles. The extensor tendons lie on the dorsal and lateral aspects of the bony elements.

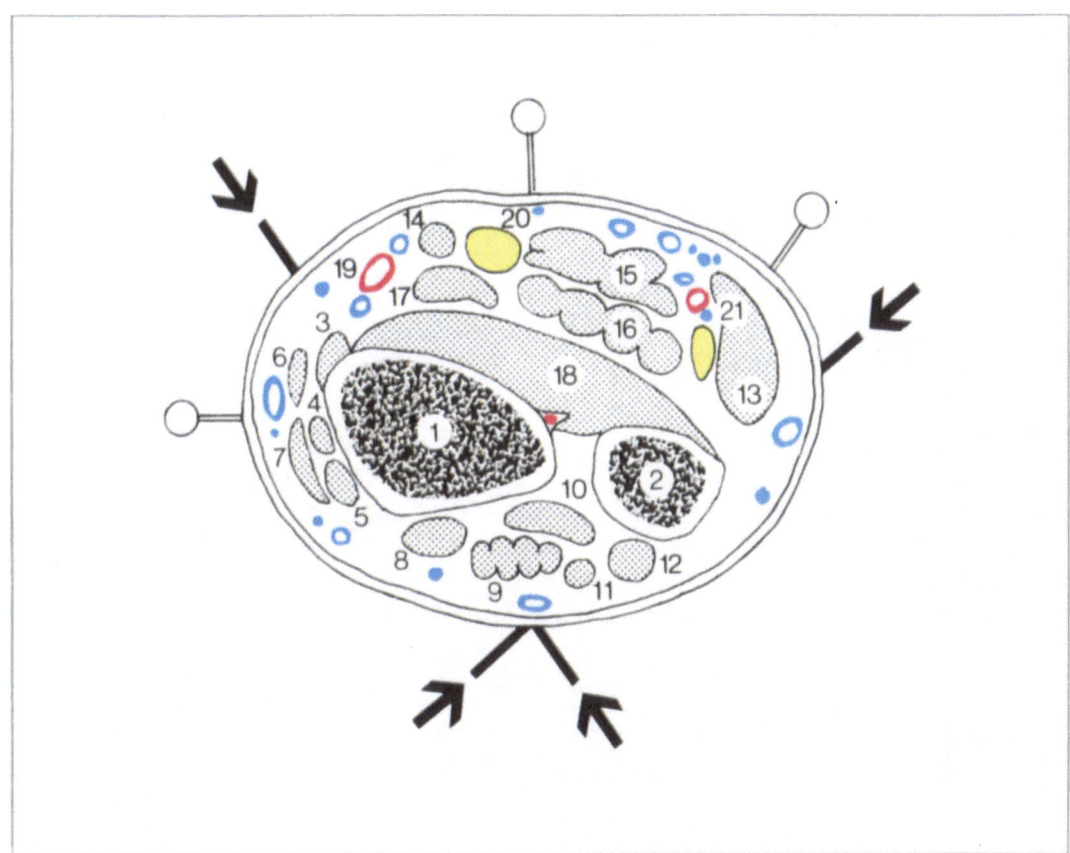

1 Radius	12 Extensor carpi ulnaris
2 Ulna	13 Flexor carpi ulnaris
3 Brachioradialis	14 Flexor carpi radialis
4 Extensor carpi radialis longus	15 Flexor digitorum superficialis
5 Extensor carpi radialis brevis	16 Flexor digitorum profundus
6 Abductor pollicis longus	17 Flexor pollicis longus
7 Extensor pollicis brevis	18 Pronator quadratus
8 Extensor pollicis longus	19 Radial vessels
9 Extensor digitorum communis	20 Median nerve
10 Extensor indicis	21 Ulnar vessels + ulnar nerve
11 Extensor digiti minimi	

Safe Areas. There is only room for a single external fixation wire in each bone at this level.

Sites for Separate External Fixation of the Ulna. These are found on the dorsal border of the tendon of the flexor carpi ulnaris muscle and the mid-dorsal aspect of the forearm.

Sites for Separate External Fixation of the Radius. These are found on the ventro-lateral edge of the distal radial metaphysis and the mid-dorsal aspect of the forearm.

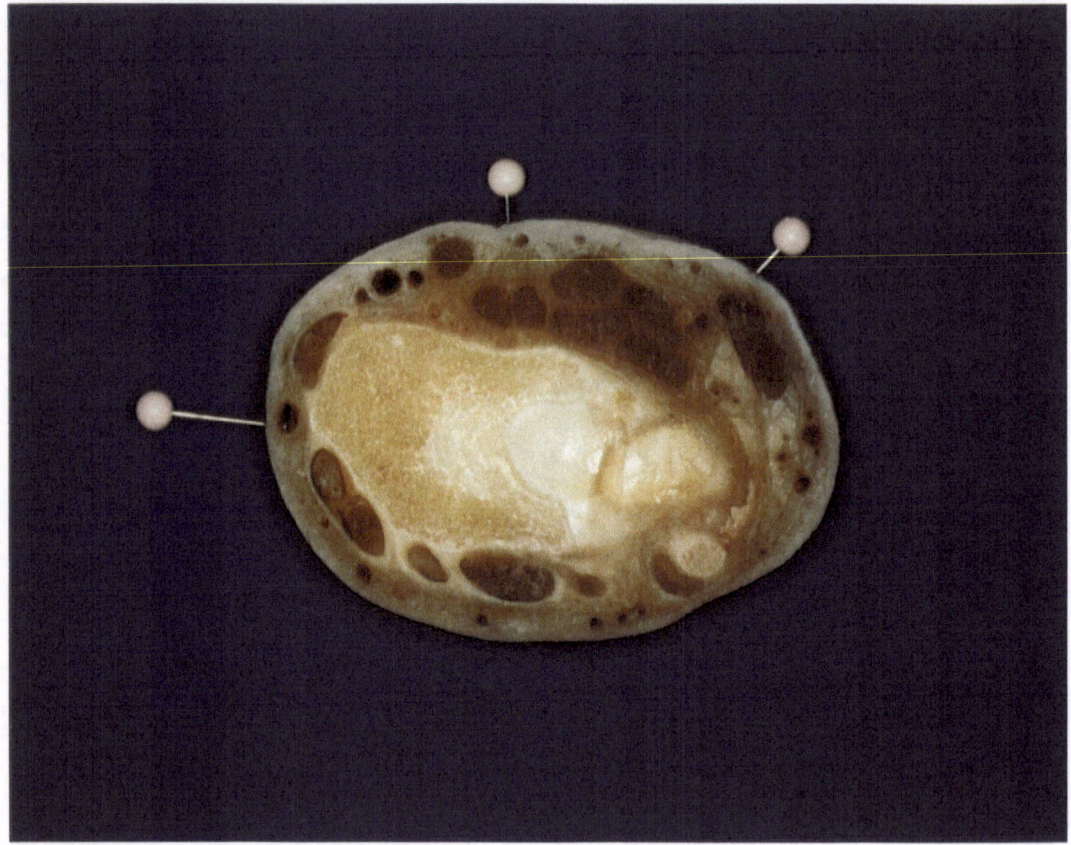

Cross-section B 8

Bones. The section has been taken through the distal radial epiphysis and shows the distal radio-ulnar and radio-carpal joints. The radius has expanded considerably, whereas the ulna now consists solely of its head and styloid process.

Vessels and Nerves. These are grouped together and lie in contact with the bony elements. The presence of the joints permits external fixation of the radius only.

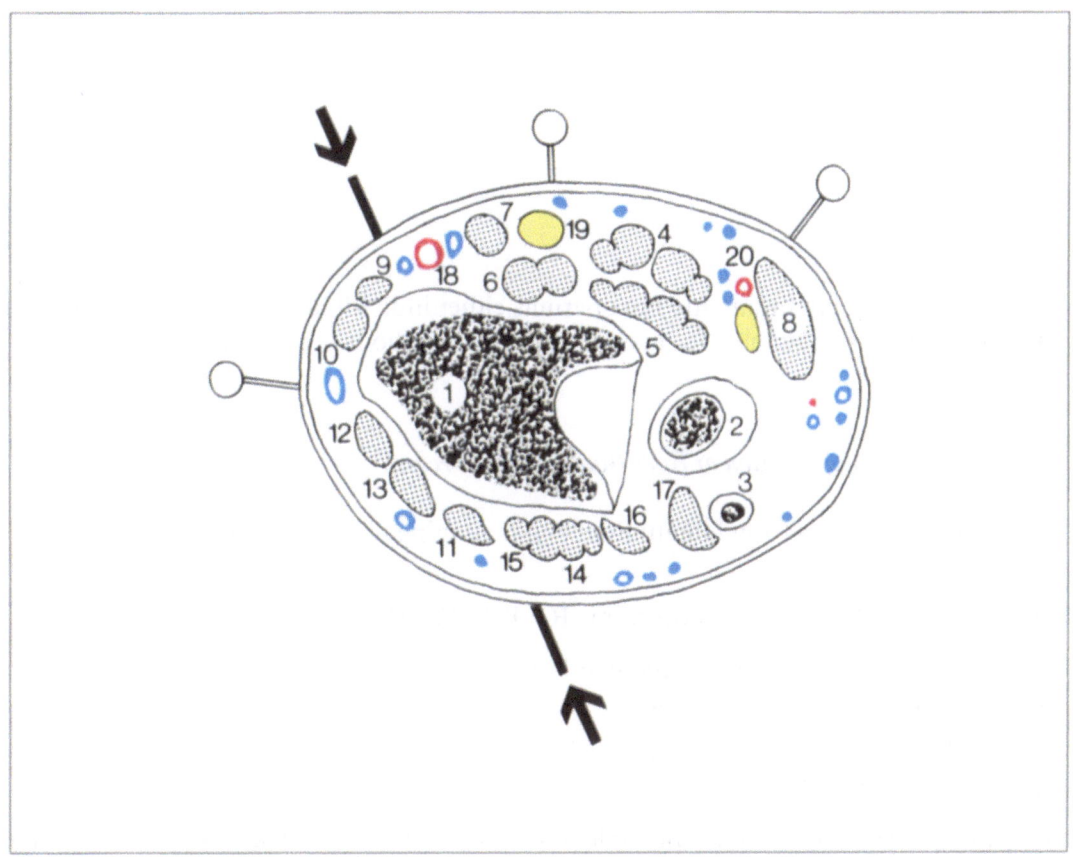

1 Radius	11 Extensor pollicis longus
2 Ulna (head)	12 Extensor carpi radialis longus
3 Ulna (styloid process)	13 Extensor carpi radialis brevis
4 Flexor digitorum superficialis	14 Extensor digitorum communis
5 Flexor digitorum profondus	15 Extensor indicis
6 Flexor pollicis longus	16 Extensor digiti minimi
7 Flexor carpi radialis	17 Extensor carpi ulnaris
8 Flexor carpi ulnaris	18 Radial vessels
9 Abductor pollicis longus	19 Median nerve
10 Extensor pollicis brevis	20 Ulnar vessels + ulnar nerve

Safe Areas. External fixation is only possible with a single wire in the radius from the ventro-lateral border to the dorsal surface.

Sites for External Fixation of the Radius. These are found on the ventro-lateral border of the distal radial epiphysis and the middle of the dorsal aspect of the forearm lateral to the tendon of the extensor digitorum communis muscle.

Safe Zones of the Forearm

The mobility of both radius and ulna permits either individual or combined transfixation. However, the zones suitable for each of these two methods are not the same.

Combined External Fixation of Both Bones of the Forearm

The cutaneous zones form longitudinal bands running along the forearm.

Proximal and Distal Epiphysis (Sections B1, B2; B7, B8)

Combined external fixation of the two bones in these regions is not possible due to the proximity of the radio-ulnar joints.

Diaphysis (Sections B3–B8)

Combined external fixation is possible in this region and lateral access is via a longitudinal cutaneous band lying ventrally and dorsally in relation to the lateral reference line. This band comprises a quarter of the region between the ventral and lateral reference lines and faces the lateral aspect of the radius.
Medial access is via a longitudinal cutaneous band lying dorso-medially in relation to the reference lines. This band is narrow and comprises a quarter of the region between the medial reference line and the dorsal border of the ulna.

The safe zones for combined transfixion of the two bones of the forearm have been named the lateral safe zone (band 1), passing from the ventral to the dorsal aspect of the forearm, and the dorso-medial safe zone, passing along the dorsal border of the ulna (band 2).

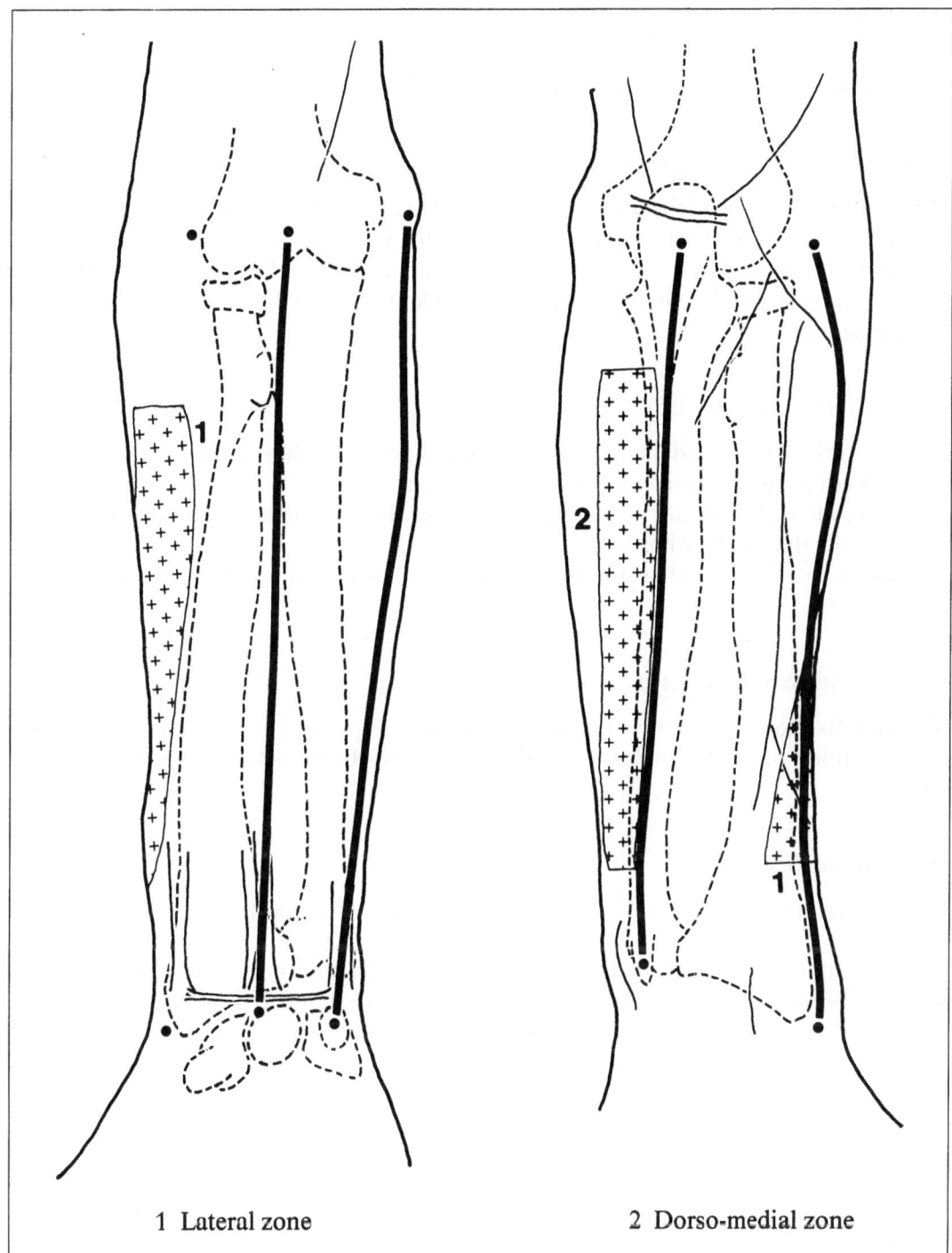

1 Lateral zone   2 Dorso-medial zone

Separate External Fixation of the Ulna

The cutaneous zones form longitudinal bands running along the forearm.

Proximal Epiphysis (Sections B 1, B 2)

Extra-articular external fixation is possible in the olecranon via dorso-lateral and dorso-medial bands related to the reference lines and the dorsal edge of the ulna. These bands represent the ventral half of the area between the medial and lateral reference lines and the dorsal border of the ulna. External fixation should always be performed dorsal to the reference lines.

Diaphysis (Sections B 3–B 6)

Access to the medial and lateral aspects of the ulna is possible via the safe zones found dorso-medial and dorso-lateral to the reference lines.
The dorso-lateral band lies in the middle third of the area between the lateral reference line and the dorsal ulnar border.
The dorso-medial band abuts the medial reference line dorsally without actually crossing it.

Distal Epiphysis (Section B 7)

External fixation is safe only when using a single wire passing between the dorsal border of the tendon of the flexor carpi ulnaris muscle and the mid-dorsal aspect of the forearm.

Ulnar Head (Section B 8)

External fixation is not possible here, as this structure is intra-articular.

The safe zones for separate external fixation of the ulna have been designated the dorso-lateral safe zone (band 1) and the dorso-medial safe zone (band 2). Both run parallel to the ulna.

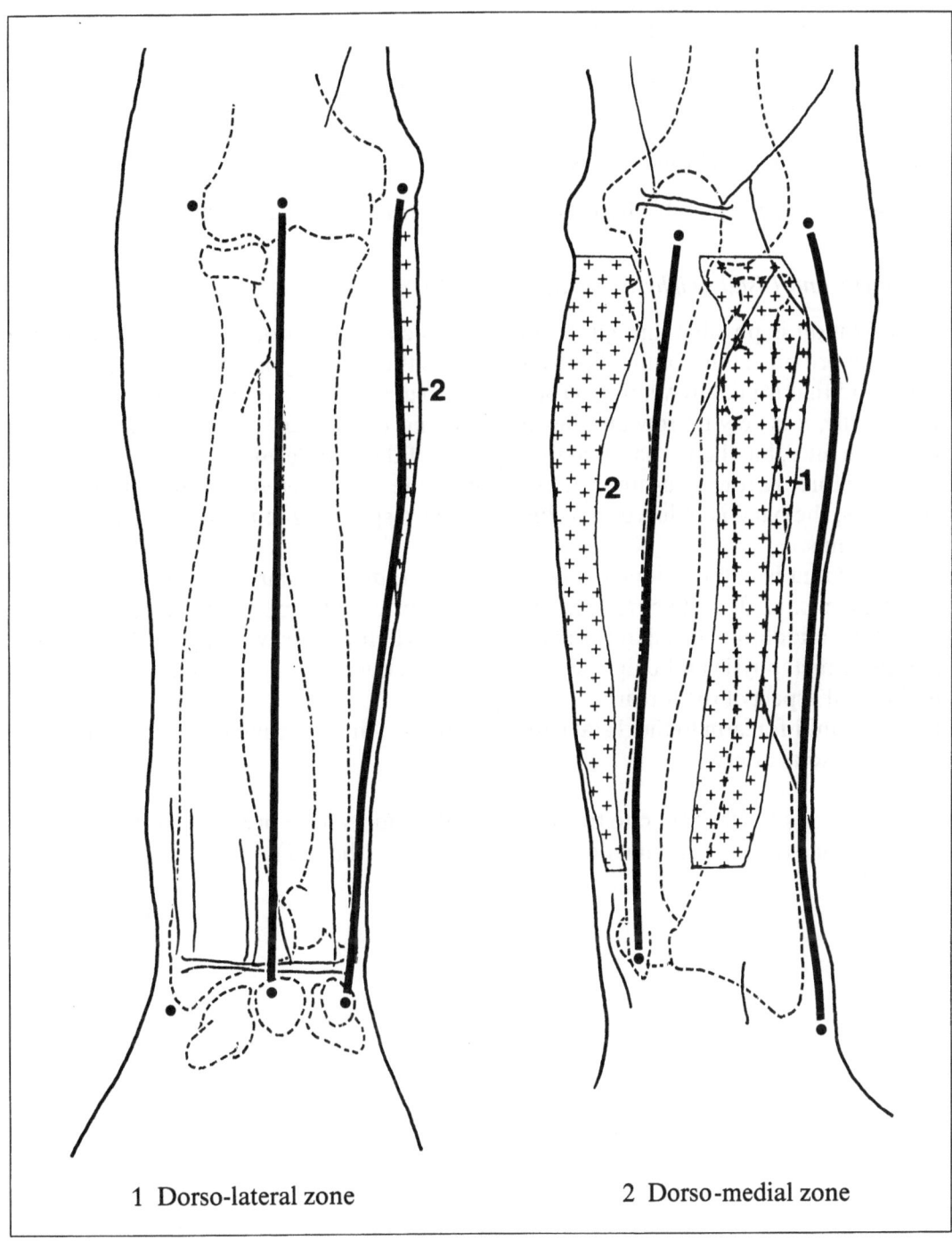

1 Dorso-lateral zone   2 Dorso-medial zone

Separate External Fixation of the Radius

The cutaneous zones are seen as longitudinal bands on the forearm.

Radial Head and Neck (Sections B 1, B 2)

Transfixation is not possible here as these structures are intra-articular.

Diaphysis and Distal Epiphysis (Sections B 3 – B 8)

Transfixation is possible through two bands which change orientation as they descend from the proximal third of the diaphysis towards the metaphysis.

The band that is initially ventral *(1)* becomes progressively lateral in relation to the bone. Proximally, it is very narrow and lies between and equidistant to the ventral and lateral reference lines, but then expands at the junction of the middle and distal thirds of the diaphysis before narrowing again as it approaches the lateral reference line. At the level of epiphysis, the band is reduced to a single line corresponding to the ventro-lateral border of the radius.

The band that is initially dorsal in relation to the radius *(2)* starts as a very narrow area, which then expands as it approaches the dorsal border of the ulna at the level of the junction of the middle and distal thirds of the diaphysis before narrowing again. At the level of the distal metaphysis and epiphysis, the band is reduced to a single line situated at the mid-dorsal aspect of the forearm.

From the mid-forearm to the distal end of the radius, the two cutaneous bands run parallel to the radial artery.

The safe zones for the separate transfixion of the radius have been called the ventro-lateral safe zone (band 1) and the dorso-lateral safe zone (band 2).

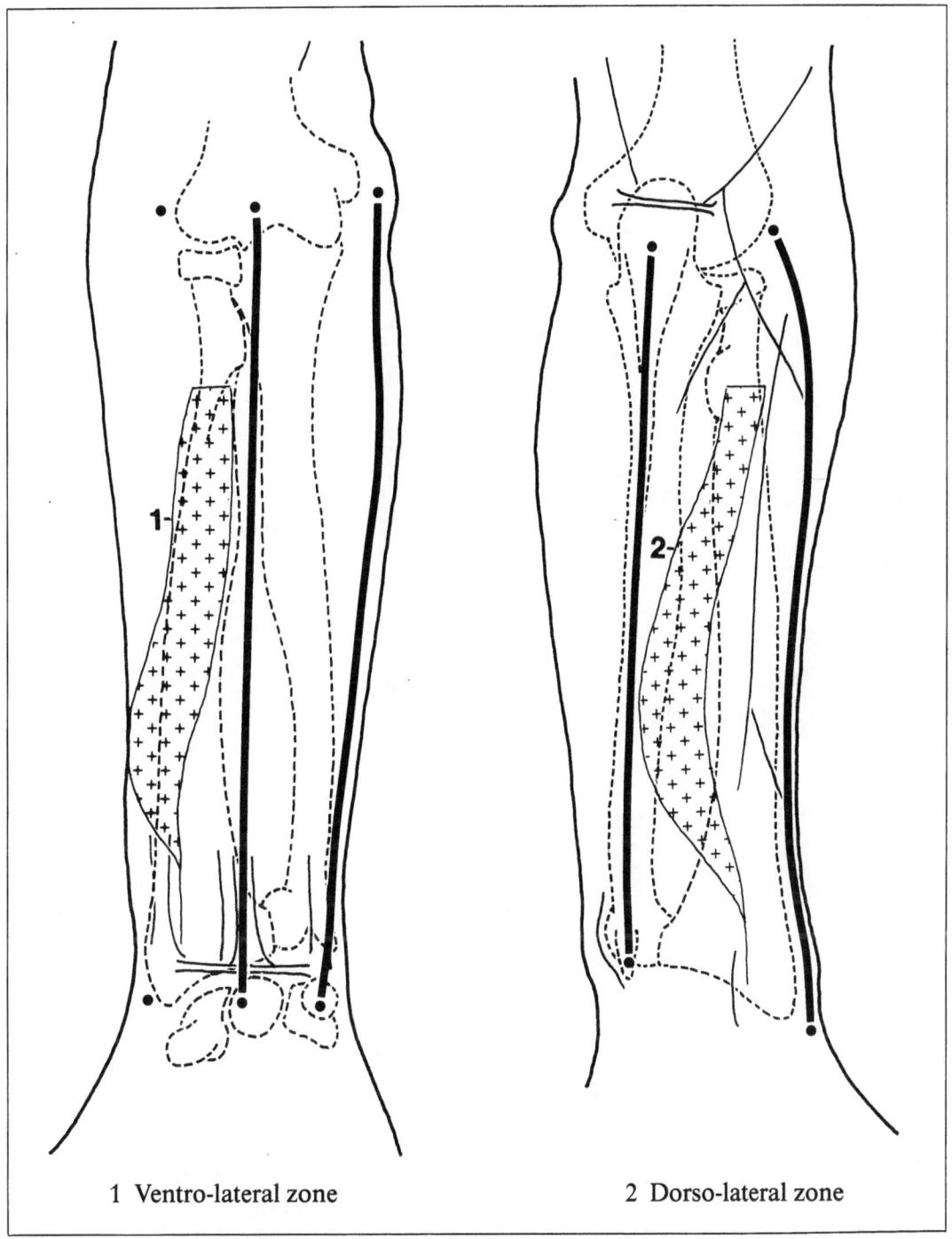

1 Ventro-lateral zone   2 Dorso-lateral zone

Computerised Tomographic Scans of the Forearm

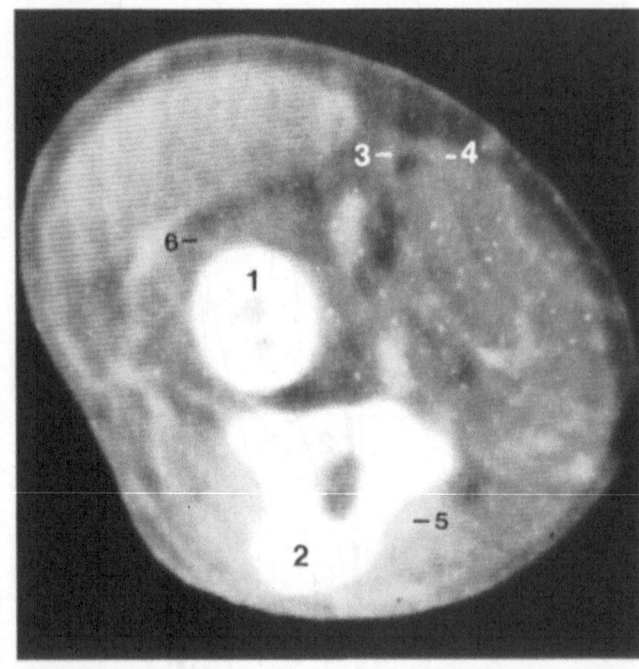

Proximal metaphysis

1 Radius (neck)
2 Ulna (olecranon)
3 Brachial vessels
4 Median nerve
5 Ulnar nerve
6 Radial nerve

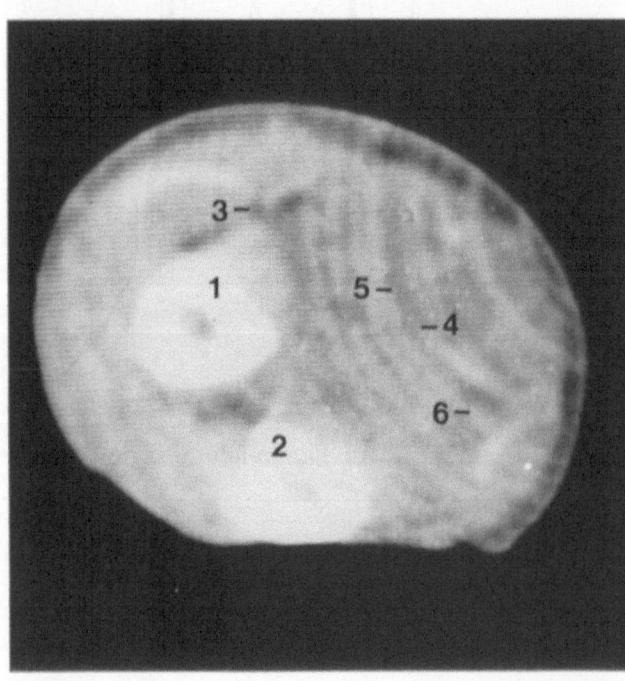

Proximal diaphysis

1 Radius
2 Ulna
3 Radial vessels
4 Ulnar vessels
5 Median nerve
6 Ulnar nerve

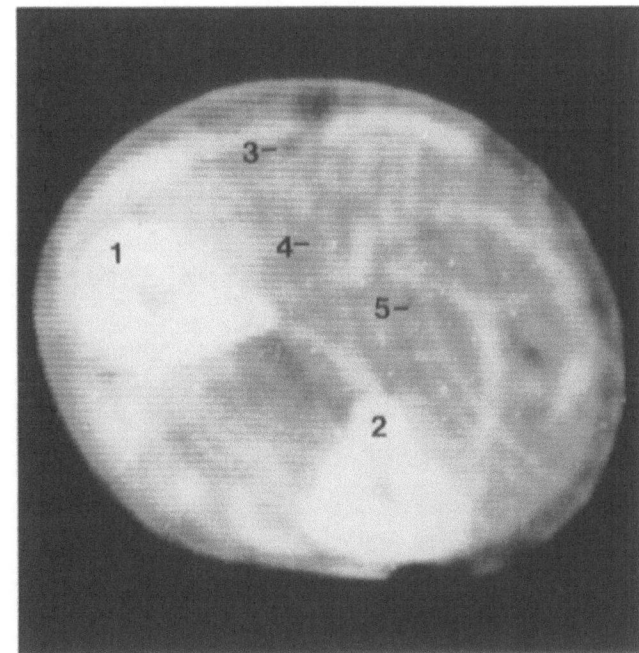

Distal diaphysis

1 Radius
2 Ulna
3 Radial vessels
4 Median nerve
5 Ulnar vessels and nerve

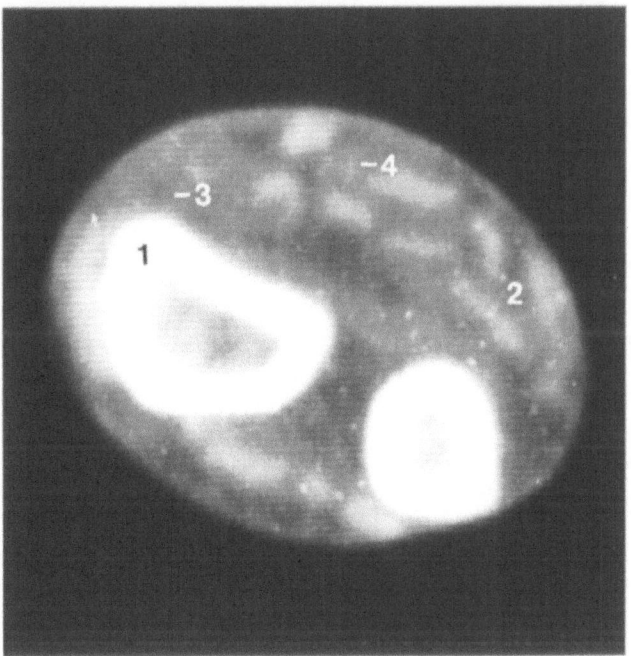

Distal metaphysis

1 Radius
2 Ulnar vessels and nerve
3 Radial vessels
4 Median nerve

C. Cross-sections of the Thigh

Levels of the Cross-sections

The 11 cross-sections are arranged in six groups corresponding to areas commonly used for external fixation.

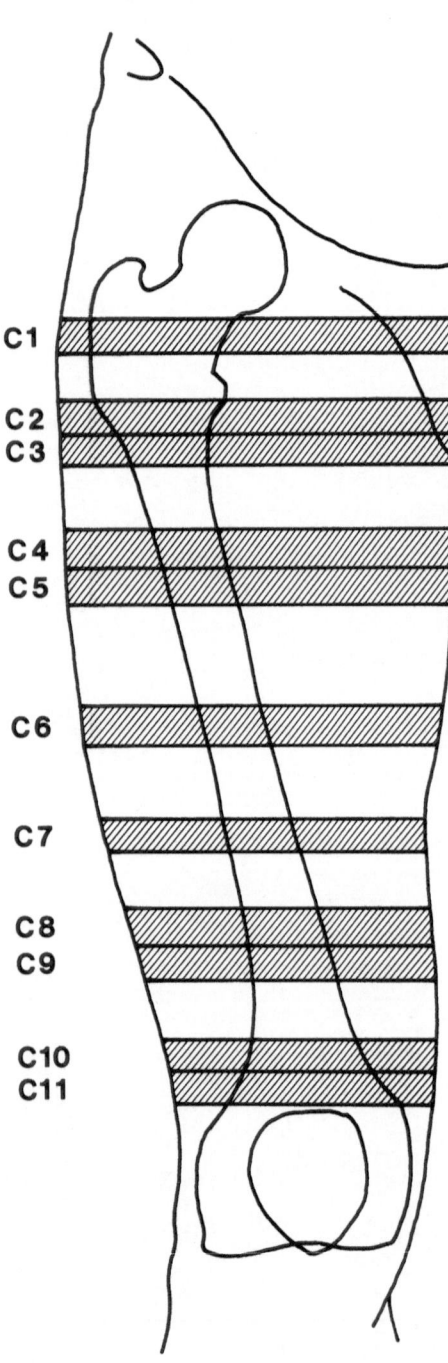

Proximal Epiphysis (C1)

The section has been taken at the level of the femoral neck and greater trochanter.

Proximal Metaphysis (C2, C3)

The sections have been taken on either side of the lesser trochanter.

Proximal Diaphysis (C4, C5)

The sections have been taken on either side of the junction of the proximal and middle thirds of the diaphysis.

Middle Diaphysis (C6, C7)

The sections have been taken on either side of a line passing through the middle of the diaphysis.

Distal Diaphysis (C8, C9)

The sections have been taken on either side of the junction of the middle and distal thirds of the diaphysis.

Distal Metaphysis (C10, C11)

The sections have been taken proximal to the superior border of the patella.

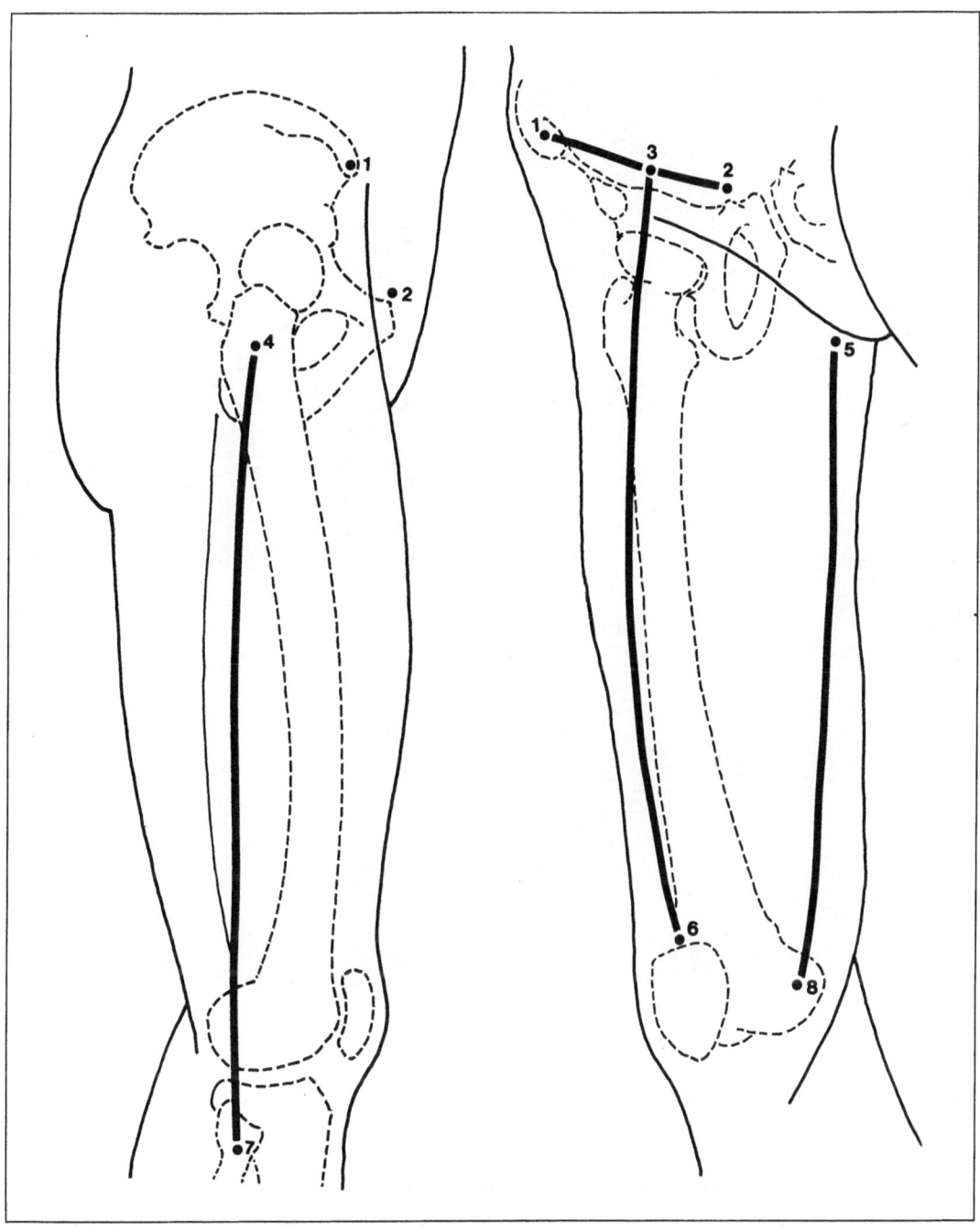

Landmarks

1 Anterior-superior iliac spine
2 Pubic tubercle
3 Midpoint of line 1–2
4 Middle of greater trochanter
5 Origin of gracilis muscle
6 Middle of the superior border of patella
7 Middle of the fibular head
8 Middle of the medial femoral condyle

Reference lines

Ventral line: line 3–6
Lateral line: line 4–7
Medial line: line 5–8

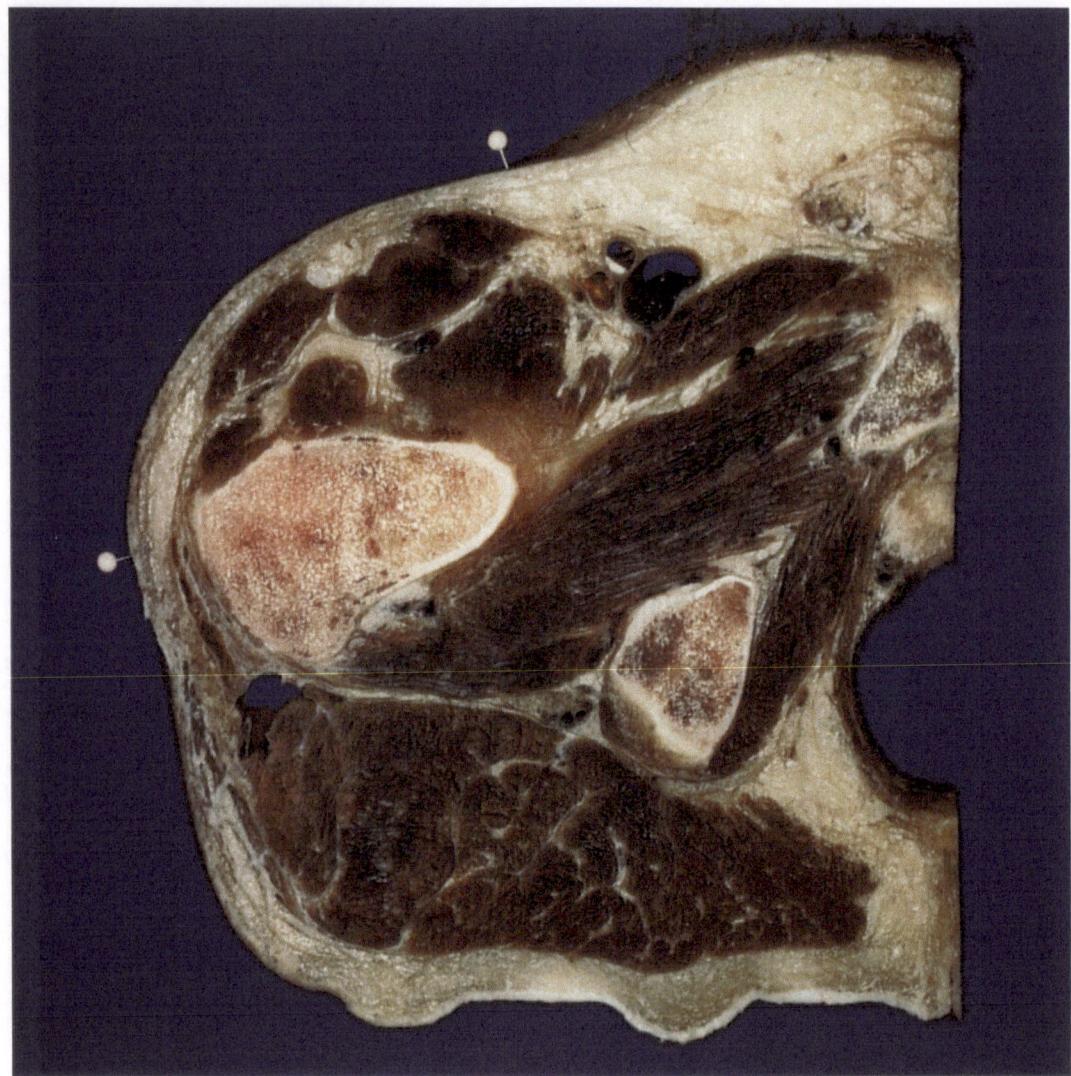

Cross-section C1

Bone. The section shows the femur (neck and greater trochanter) and the pelvis (ischium and pubis).

Vessels and Nerves. The femoral artery and vein are ventro-medial in relation to the femur. The femoral nerve and its branches lie between the vessels and the sartorius muscle. The sciatic nerve lies opposite the femoral vessels in the sagittal plane.

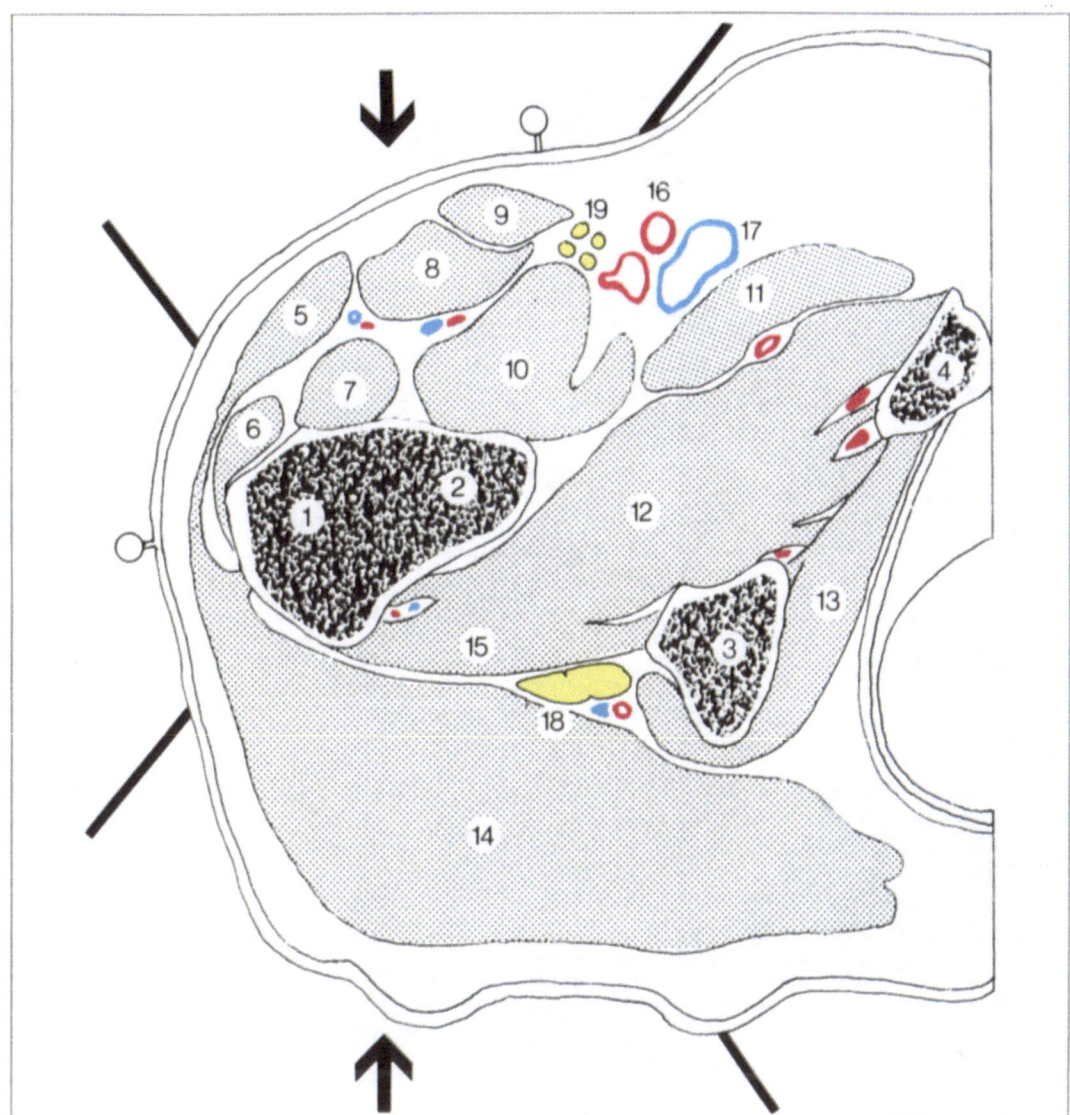

1 Femur (greater trochanter)	8 Rectus femoris	14 Gluteus maximus
2 Femur (neck)	9 Sartorius	15 Quadratus femoris
3 Ischium	10 Iliopsoas	16 Femoral artery
4 Pubis	11 Pectineus	17 Femoral vein
5 Tensor fasciae latae	12 Obturator externus	18 Sciatic nerve
6 Vastus lateralis	13 Obturator internus	19 Femoral nerve
7 Vastus intermedius		

Safe Areas. External fixation is not possible in the direction of the pelvis. The safe areas are extensive here and lie ventrally and dorsally in relation to the femur.

Cutaneous Zones Related to the Safe Areas. These comprise two zones lying ventro-laterally and dorso-laterally in relation to the reference lines.

The ventro-lateral zone, consisting of the ventral half of the area lying between the ventral and lateral reference lines, and *the dorso-lateral zone* are symmetrically disposed about the lateral reference line.

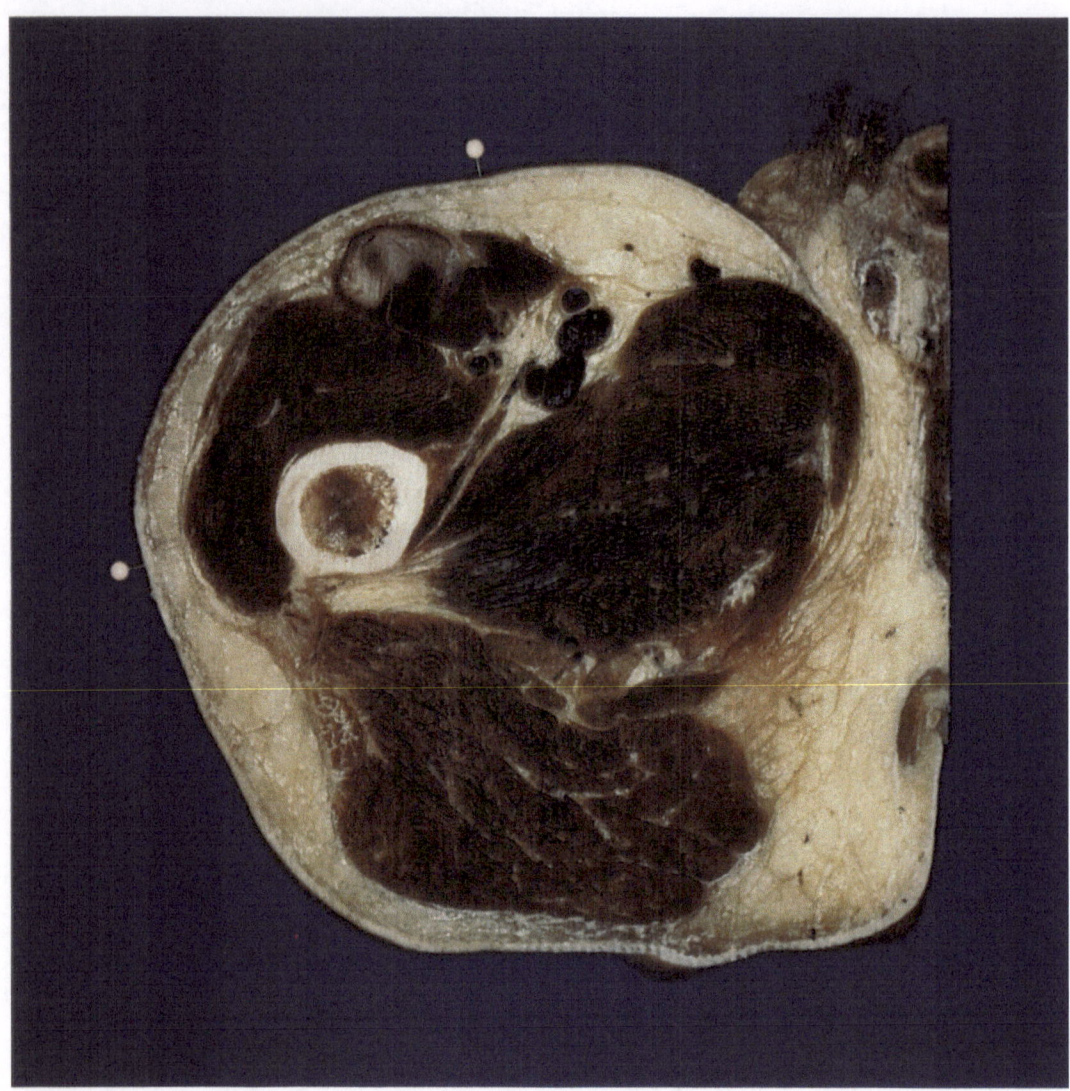

Cross-section C2

Bone. The section shows the femur at the level of the lesser trochanter.

Vessels and Nerves. The femoral vessels diverge from the profunda vessels at this level. These two groups of vessels lie ventro-medially in relation to the femur. The femoral nerve has divided into its branches. The sciatic nerve lies opposite the femoral vessels in the sagittal plane.

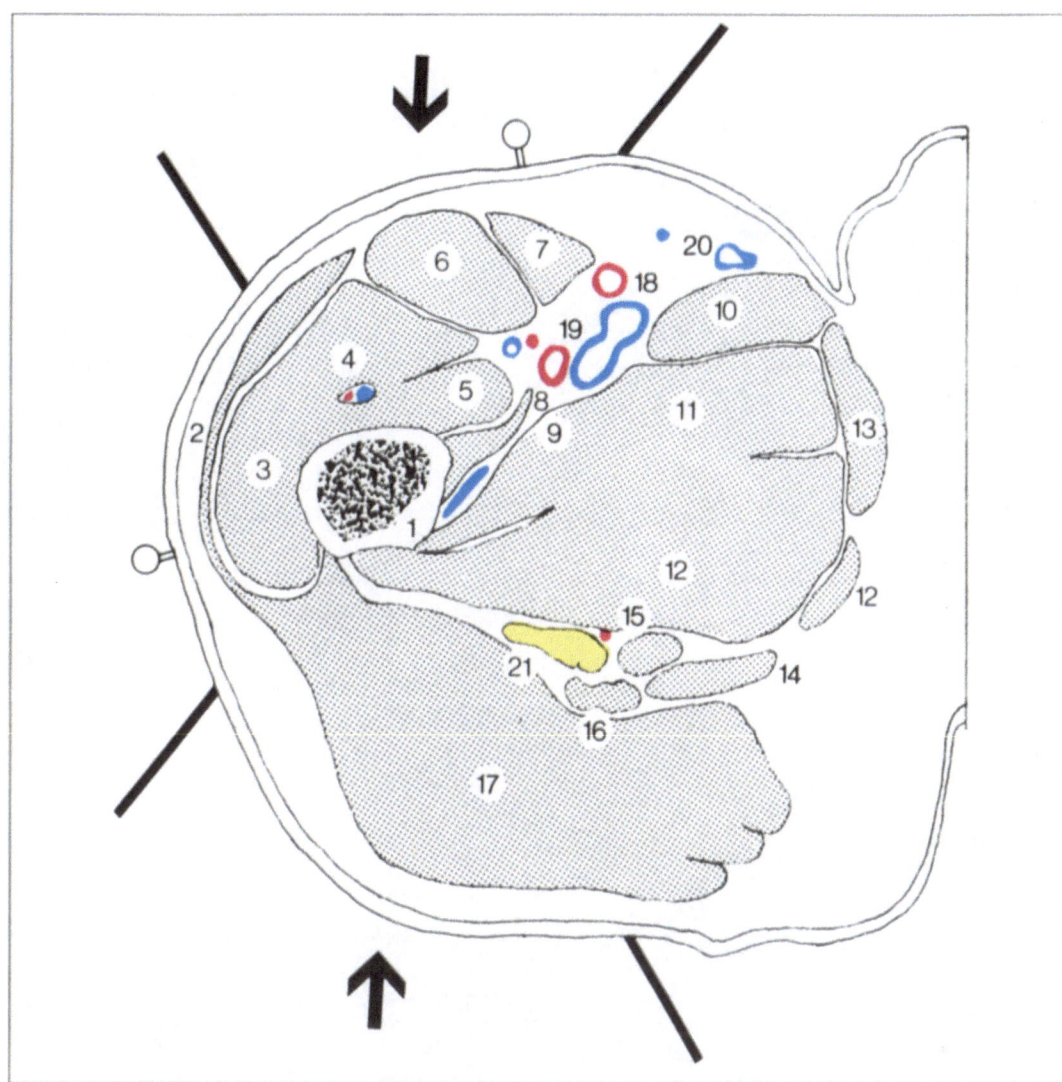

1 Femur (lesser trochanter)	8 Iliopsoas	15 Semimembranosus
2 Tensor fasciae latae	9 Pectineus	16 Biceps femoris (long head)
3 Vastus lateralis	10 Adductor longus	17 Gluteus maximus
4 Vastus intermedius	11 Adductor brevis	18 Femoral vessels
5 Vastus medialis	12 Adductor magnus	19 Profunda vessels
6 Rectus femoris	13 Gracilis	20 Long saphenous vein
7 Sartorius	14 Semitendinosus	21 Sciatic nerve

Safe Areas. External fixation is not possible in the direction of the pelvis at this level. The safe areas are large and lie ventrally and dorsally in relation to the femur.

Cutaneous Zones Related to the Safe Areas. These comprise two zones lying ventro-laterally and dorso-laterally in relation to the reference lines.

- *The ventro-lateral zone* begins just medial to the ventral reference line and consists of the ventral half of the area between the ventral and lateral reference lines.

- *The dorso-lateral zone* lies opposite the ventro-lateral zone; the two zones are symmetrically situated with respect to the axis of the lateral reference line.

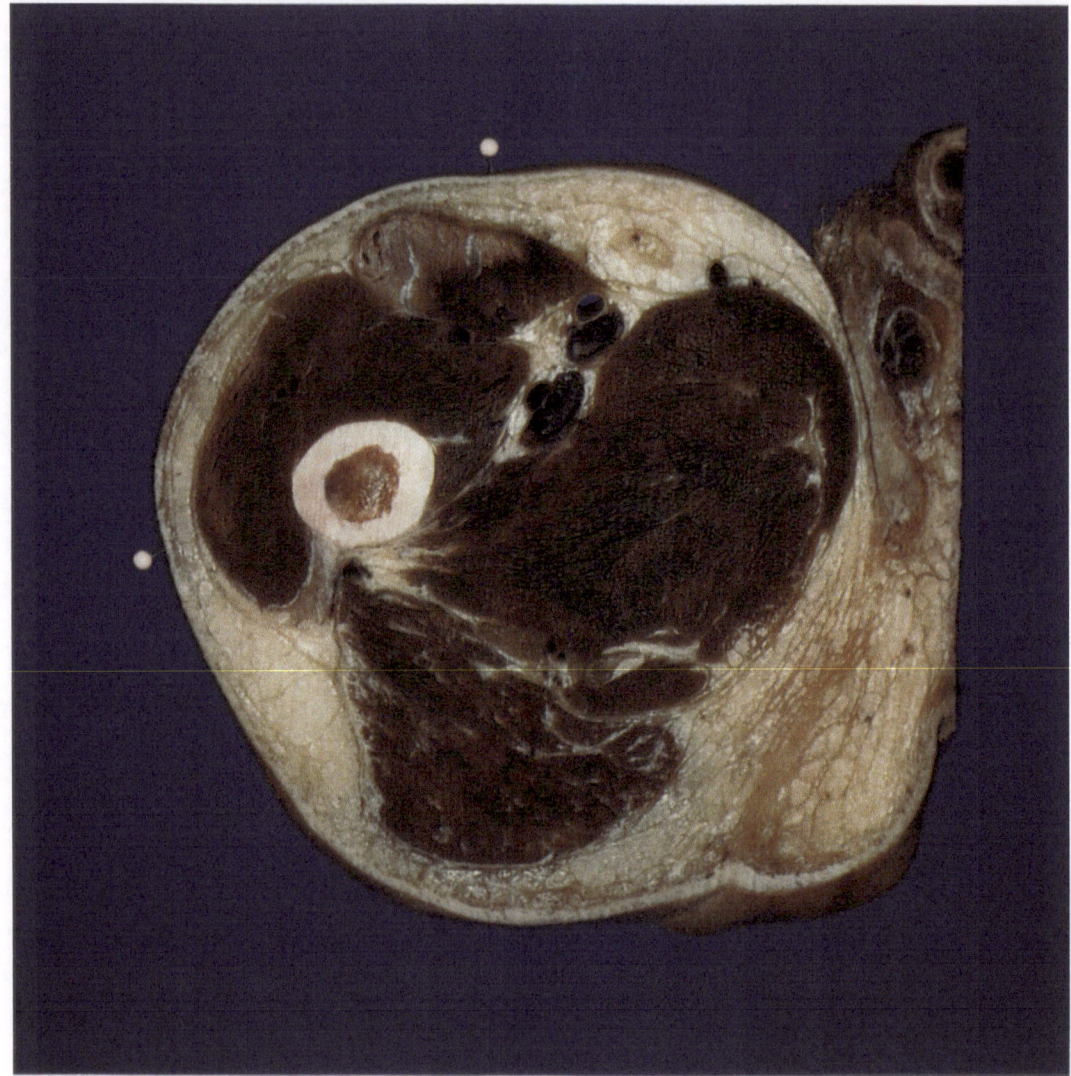

Cross-section C3

Bone. The section shows the proximal femoral metaphysis.

Vessels and Nerves. The femoral vessels are now separate from the profunda vessels, which pass towards the femur. The sciatic nerve remains opposite the femoral vessels in the sagittal plane.

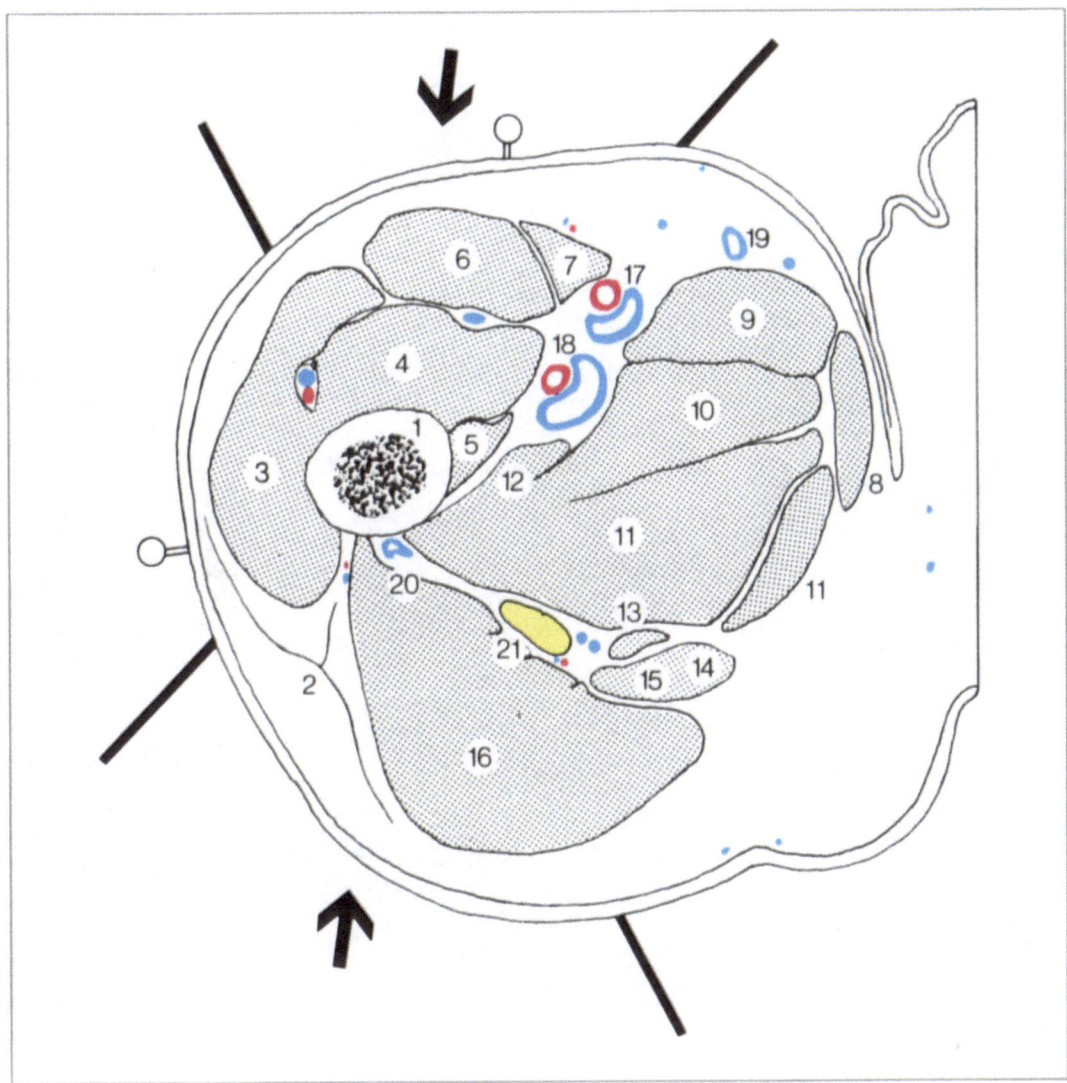

1 Femur	8 Gracilis	15 Biceps femoris (long head)
2 Iliotibial tract	9 Adductor longus	16 Gluteus maximus
3 Vastus lateralis	10 Adductor brevis	17 Femoral vessels
4 Vastus intermedius	11 Adductor magnus	18 Profunda vessels
5 Vastus medialis	12 Pectineus	19 Long saphenous vein
6 Rectus femoris	13 Semimembranosus	20 Perforating vessels
7 Sartorius	14 Semitendinosus	21 Sciatic nerve

Safe Areas. Transfixion is not possible in the direction of the perineum. The safe areas are large and lie ventrally and laterally in relation to the femur.

Cutaneous Zones Related to the Safe Areas. These comprise two zones lying ventrally and dorso-laterally in relation to the reference lines.
- *The ventral zone* crosses the ventral reference line and extends as far as the sartorius muscle. Its lateral border is located halfway between the ventral and lateral reference lines.
- *The dorso-lateral zone* is limited dorsally to the middle of the gluteus maximus muscle and laterally by the dorsal border of the vastus lateralis muscle, stopping short of the lateral reference line.

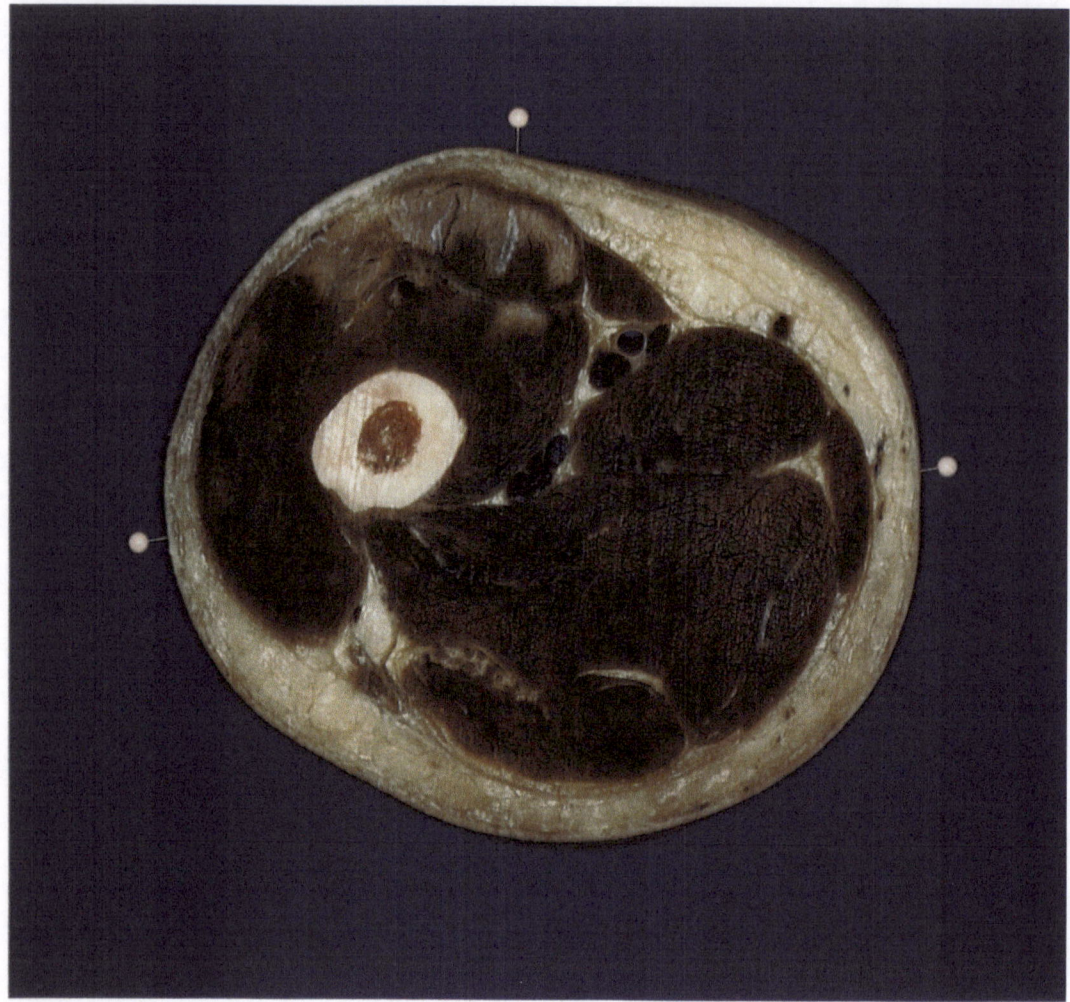

Cross-section C4

Bone. The section has been taken at the junction of the proximal and middle thirds of the diaphysis. The bone lies laterally in this circular section.

Vessels and Nerves. The femoral vessels are less ventral in relation to the femur and lie underneath the sartorius muscle. The profunda vessels pass towards the dorsal aspect of the femur. The sciatic nerve approaches the femur.

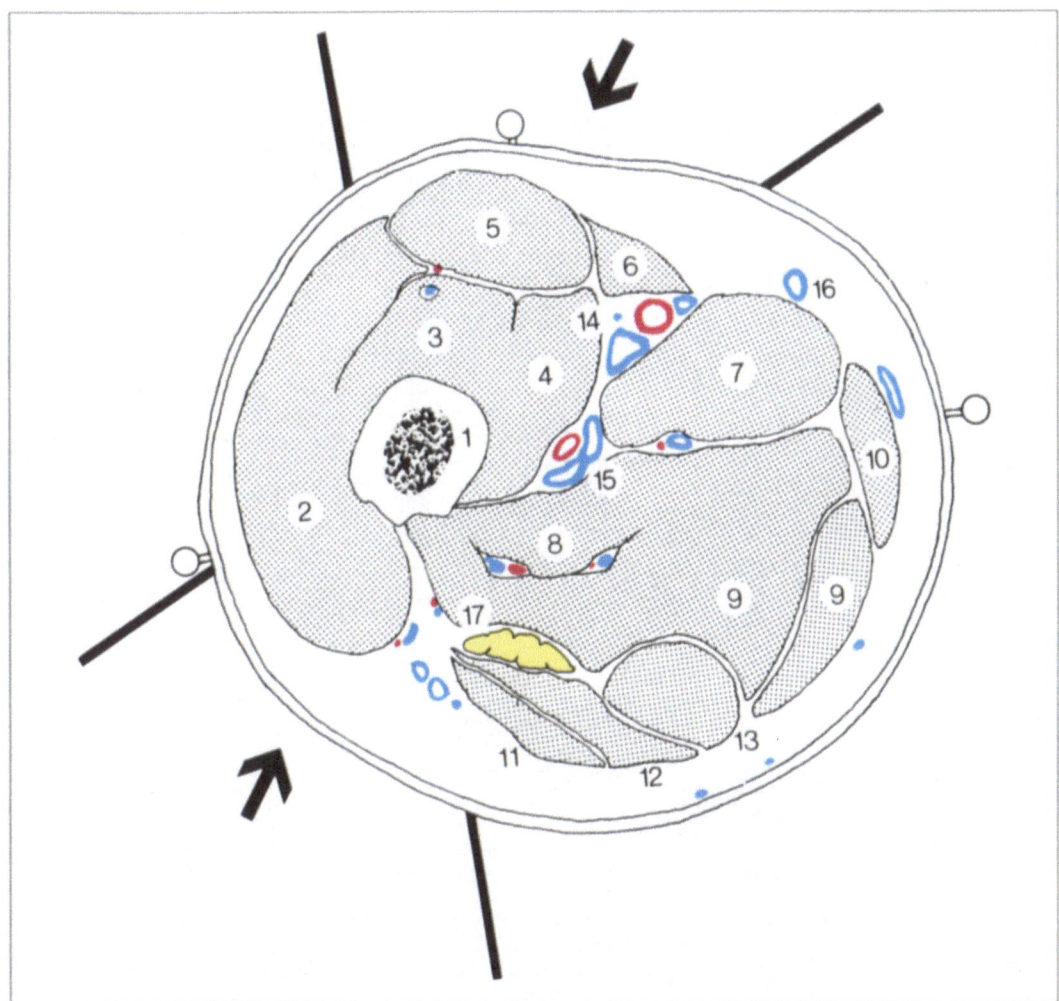

1 Femur	7 Adductor longus	13 Semimembranosus
2 Vastus lateralis	8 Adductor brevis	14 Femoral vessels
3 Vastus intermedius	9 Adductor magnus	15 Profunda vessels
4 Vastus medialis	10 Gracilis	16 Long saphenous vein
5 Rectus femoris	11 Biceps femoris (long head)	17 Sciatic nerve
6 Sartorius	12 Semitendinosus	

Safe Areas. These are extensive and lie ventro-medially and dorso-laterally in relation to the femur.

Cutaneous Zones Related to the Safe Areas. These comprise two zones lying ventro-medially and dorso-laterally in relation to the reference lines.

- *The ventro-medial zone* extends from the lateral border of the rectus femoris muscle to the middle of the area between the ventral and medial reference lines (i.e. to the body of the sartorius muscle).

- *The dorso-lateral zone* extends from the lateral reference line to the lateral border of the biceps femoris muscle.

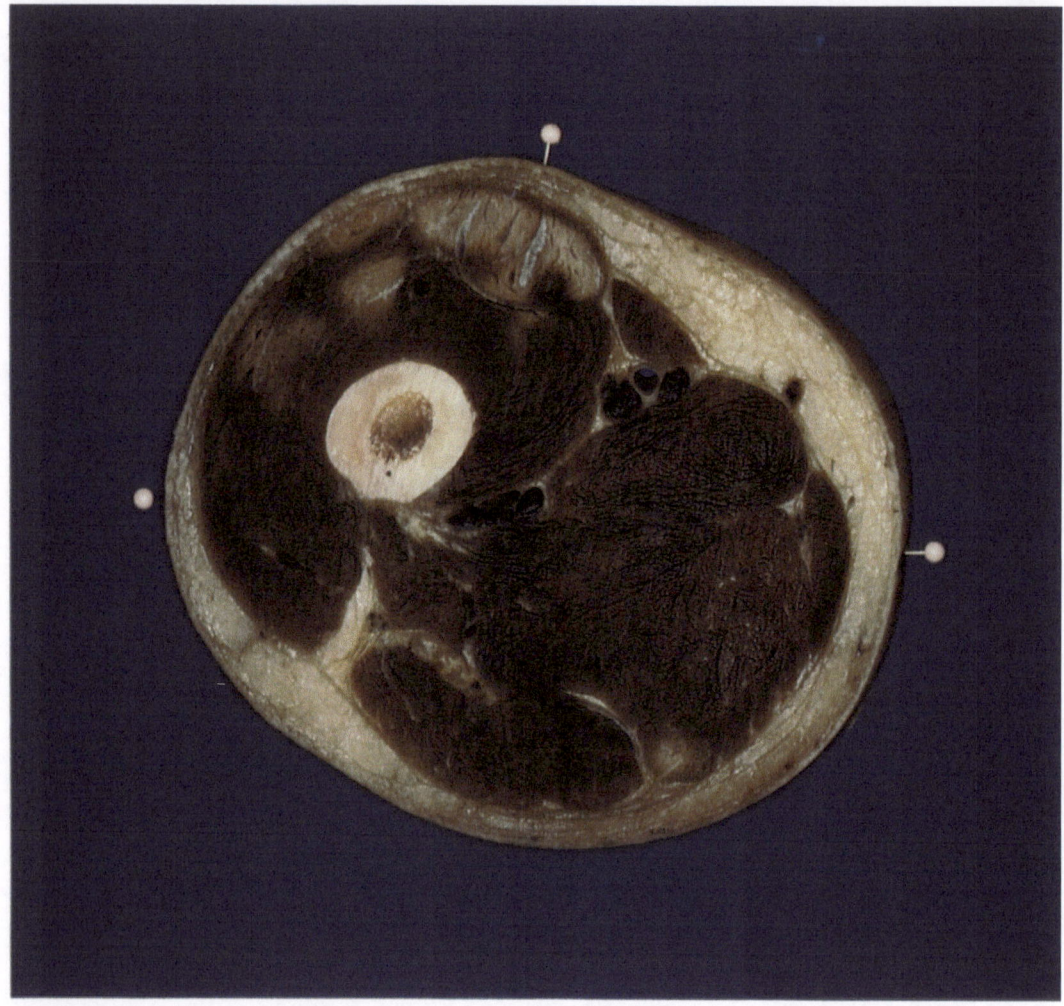

Cross-section C5

Bone. The section has been taken just distal to the junction of the proximal and middle thirds of the diaphysis. The bone lies eccentrically in a ventro-lateral direction.

Vessels and Nerves. The femoral vessels continue to pass dorsally deep to the sartorius muscle. The profunda vessels are close to the linea aspera. The sciatic nerve lies in the sagittal plane behind the femur.

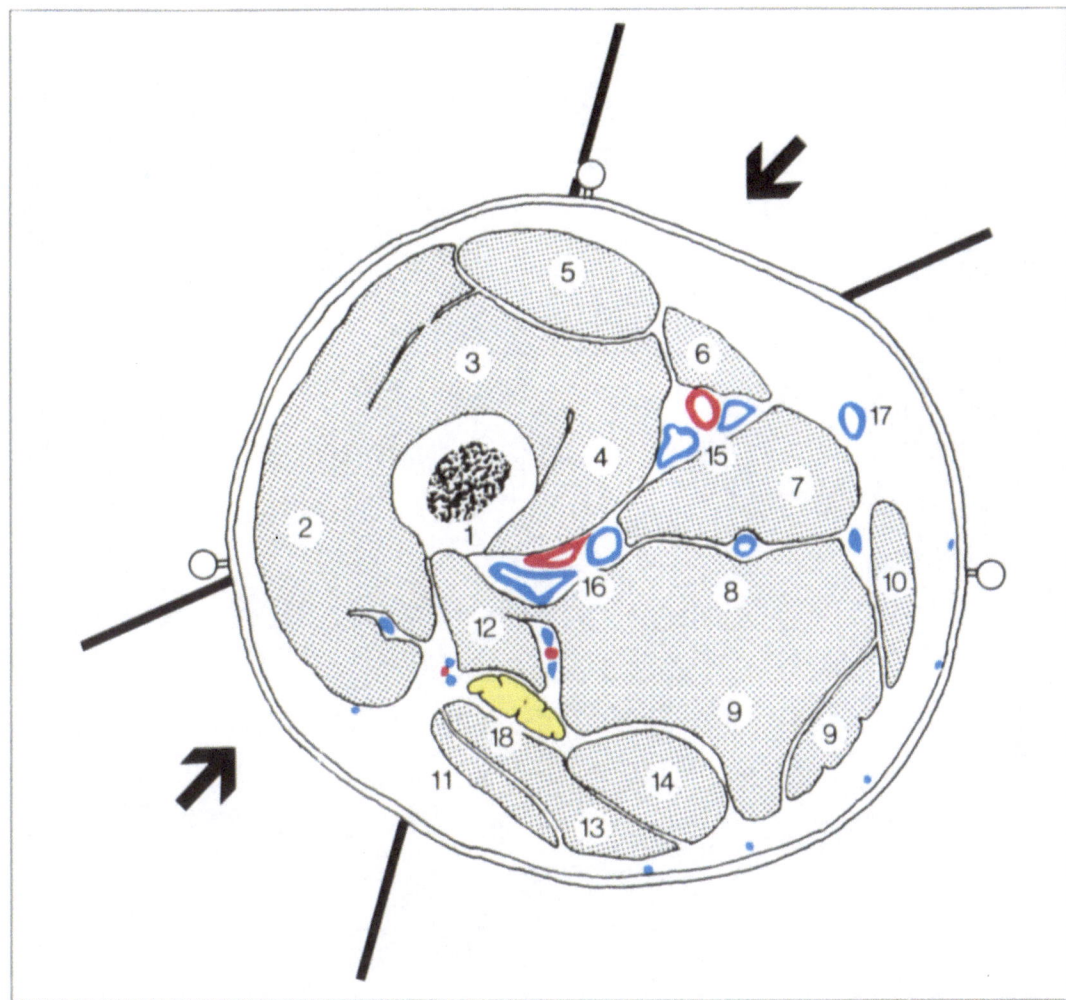

1 Femur	7 Adductor longus	13 Semitendinosus
2 Vastus lateralis	8 Adductor brevis	14 Semimembranosus
3 Vastus intermedius	9 Adductor magnus	15 Femoral vessels
4 Vastus medialis	10 Gracilis	16 Profunda vessels
5 Rectus femoris	11 Biceps femoris (long head)	17 Long saphenous vein
6 Sartorius	12 Biceps femoris (short head)	18 Sciatic nerve

Safe Areas. These are less extensive at this level and lie ventro-medially and dorso-laterally in relation to the femur.

Cutaneous Zones Related to the Safe Areas. These form two zones lying ventro-medially and dorso-laterally in relation to the reference lines.

- *The ventro-medial zone* lies in the ventral half of the area between the ventral and the medial reference lines (the latter ending at the sartorius muscle).

- *The dorso-lateral zone* lies between the lateral reference line and the dorsal border of the vastus lateralis muscle.

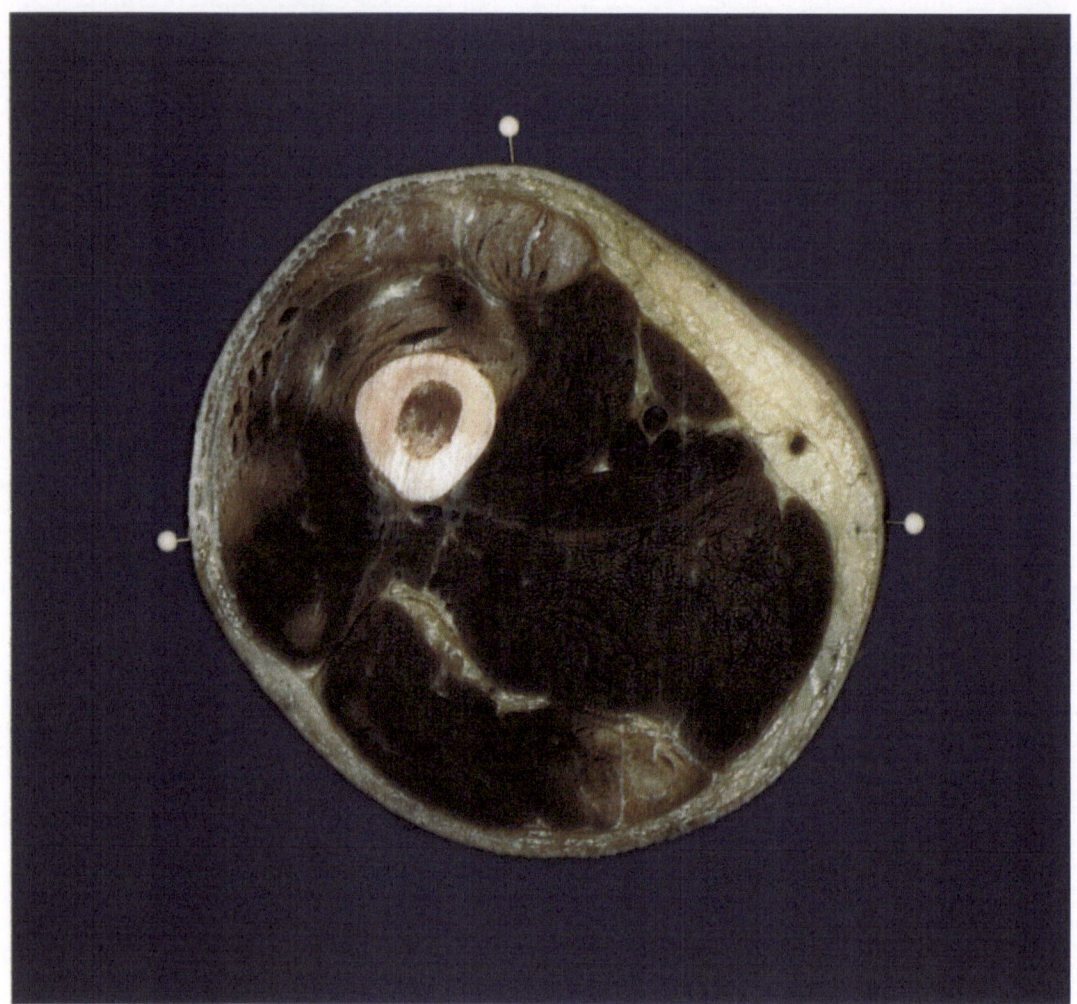

Cross-section C6

Bone. The section has been taken just proximal to the mid-point of the diaphysis. The bone lies ventro-laterally.

Vessels and Nerves. The femoral vessels lie in the same frontal plane as the femur. The profunda vessels have now divided into small branches. The sciatic nerve approaches the dorsal aspect of the femur.

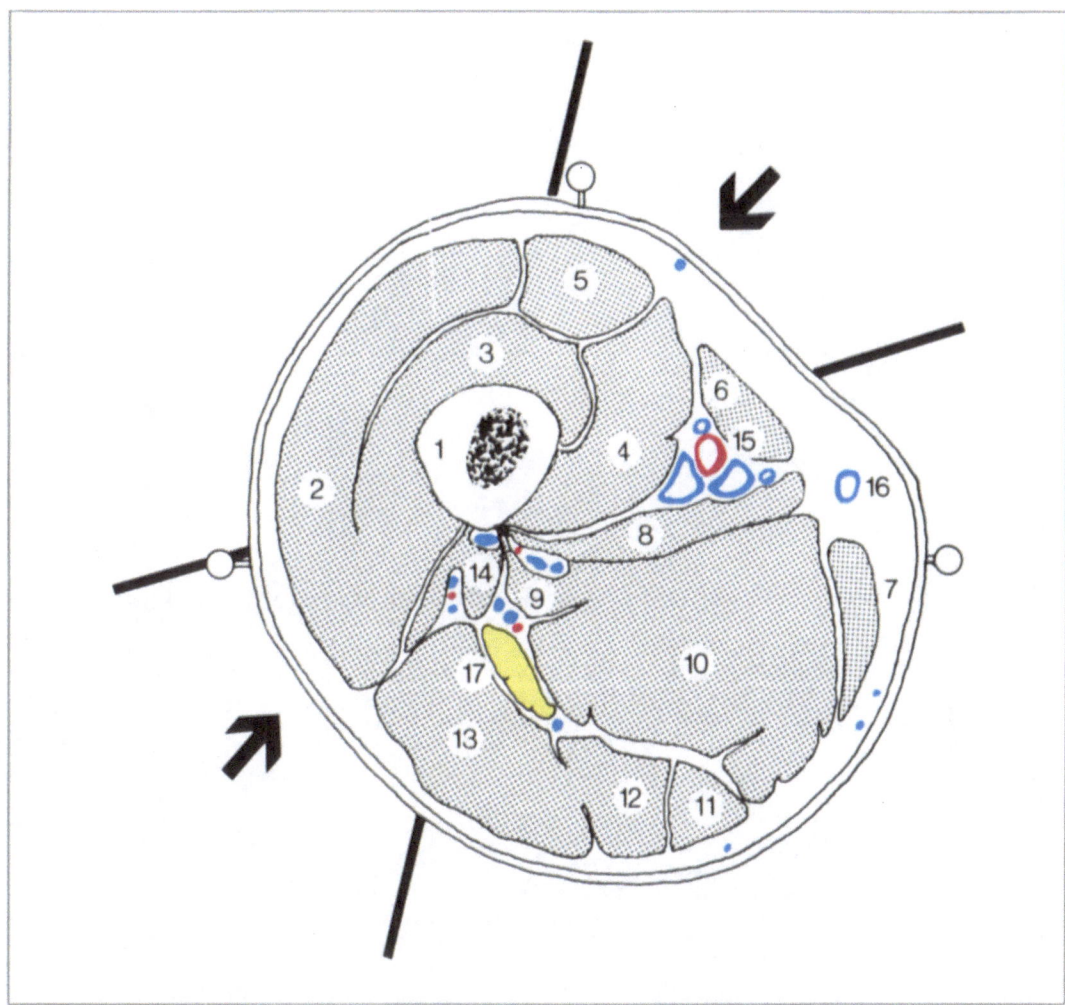

1 Femur	7 Gracilis	13 Biceps femoris (long head)
2 Vastus lateralis	8 Adductor longus	14 Biceps femoris (short head)
3 Vastus intermedius	9 Adductor brevis	15 Femoral vessels
4 Vastus medialis	10 Adductor magnus	16 Long saphenous vein
5 Rectus femoris	11 Semimembranosus	17 Sciatic nerve
6 Sartorius	12 Semitendinosus	

Safe Areas. These lie ventro-medially and dorso-laterally in relation to the femur.

Cutaneous Zones Related to the Safe Areas. These comprise two zones lying ventro-medially and dorso-laterally in relation to the reference lines.

- *The ventro-medial zone* extends from the ventral reference line to the middle of the area bounded by the ventral and the medial reference lines. The latter corresponds to the sartorius muscle.

- *The dorso-lateral zone* lies between the lateral reference line and the dorsal edge of the vastus lateralis muscle.

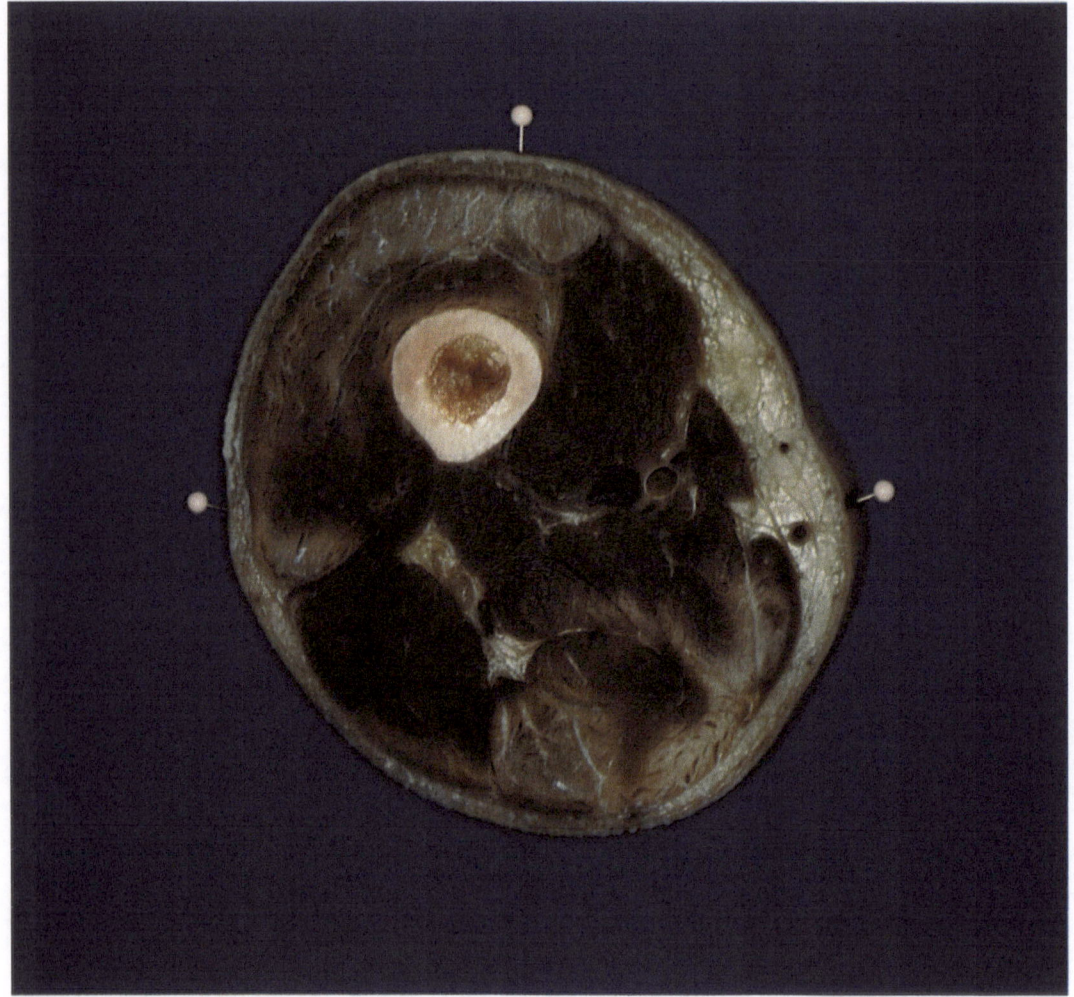

Cross-section C7

Bone. The section has been taken just distal to the middle the femoral diaphysis, the diameter of which has become larger at this level.

Vessels and Nerves. The femoral vessels lie dorso-medially in relation to the femur. The sciatic nerve lies dorsally and close to the femur.

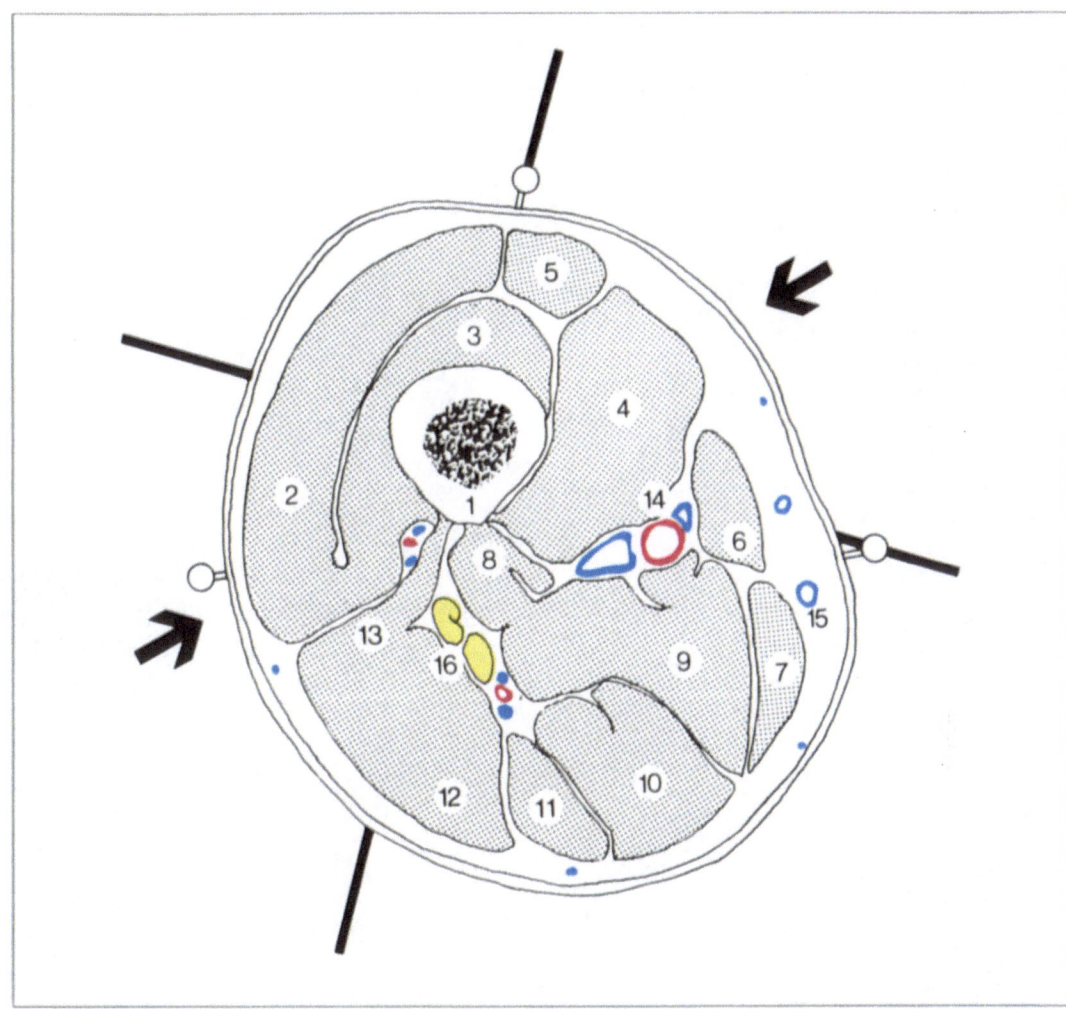

1 Femur	7 Gracilis	12 Biceps femoris (long head)
2 Vastus lateralis	8 Adductor longus	13 Biceps femoris (short head)
3 Vastus intermedius	9 Adductor magnus	14 Femoral vessels
4 Vastus medialis	10 Semimembranosus	15 Long saphenous vein
5 Rectus femoris	11 Semitendinosus	16 Sciatic nerve
6 Sartorius		

Safe Areas. These are large and lie ventro-medially and dorso-laterally in relation to the femur.

Cutaneous Zones Related to the Safe Areas. These comprise two zones lying ventro-medially and laterally in relation to the reference lines.

– *The ventro-medial zone* lies between the medial and ventral reference lines.

– *The lateral zone* extends from the middle of the area bounded by the ventral and lateral reference lines (i.e. from the middle of the body of the vastus lateralis muscle) to the middle of the body of the biceps femoris muscle.

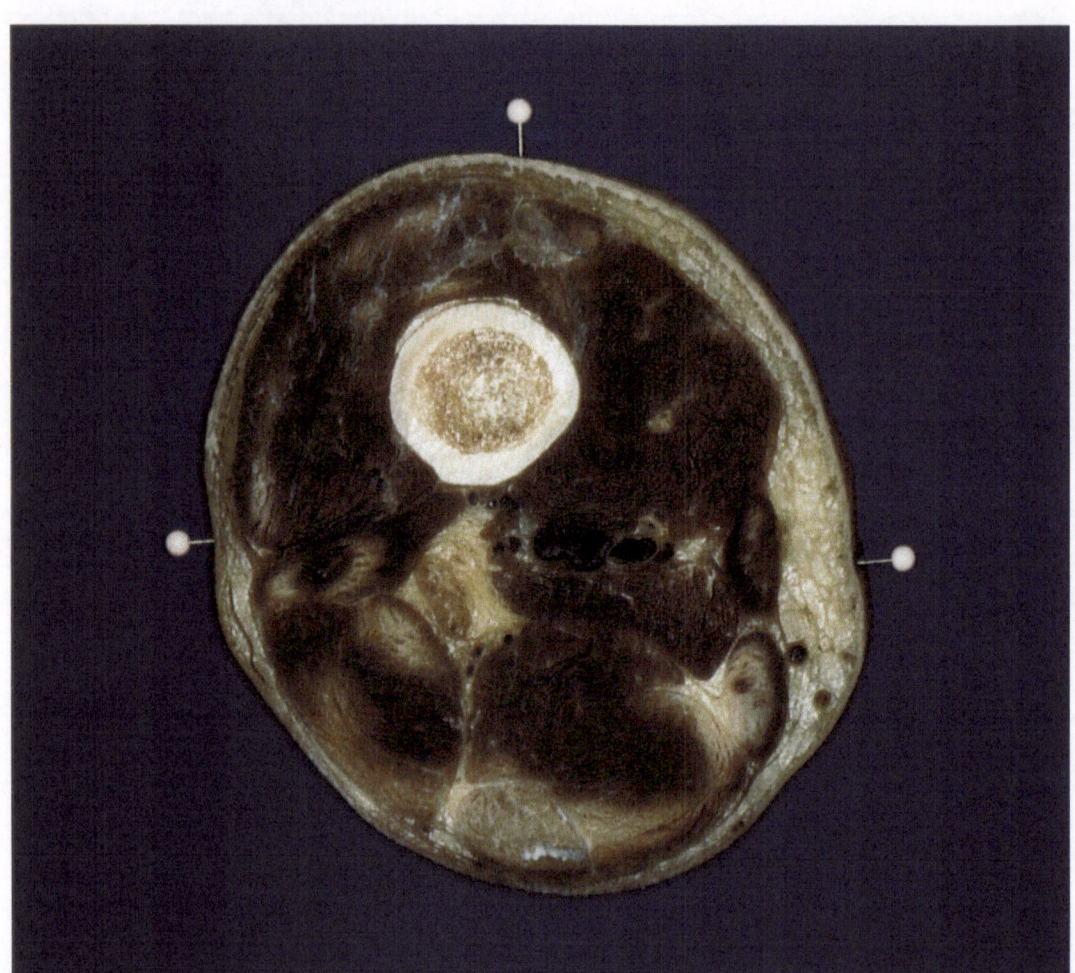

Cross-section C8

Bone. The section has been taken just proximal to the junction of the middle and distal thirds of the diaphysis. The femur, the diameter of which has increased, lies in the ventral half of the section.

Vessels and Nerves. The femoral vessels and the sciatic nerve lie dorsally in relation to the femur.

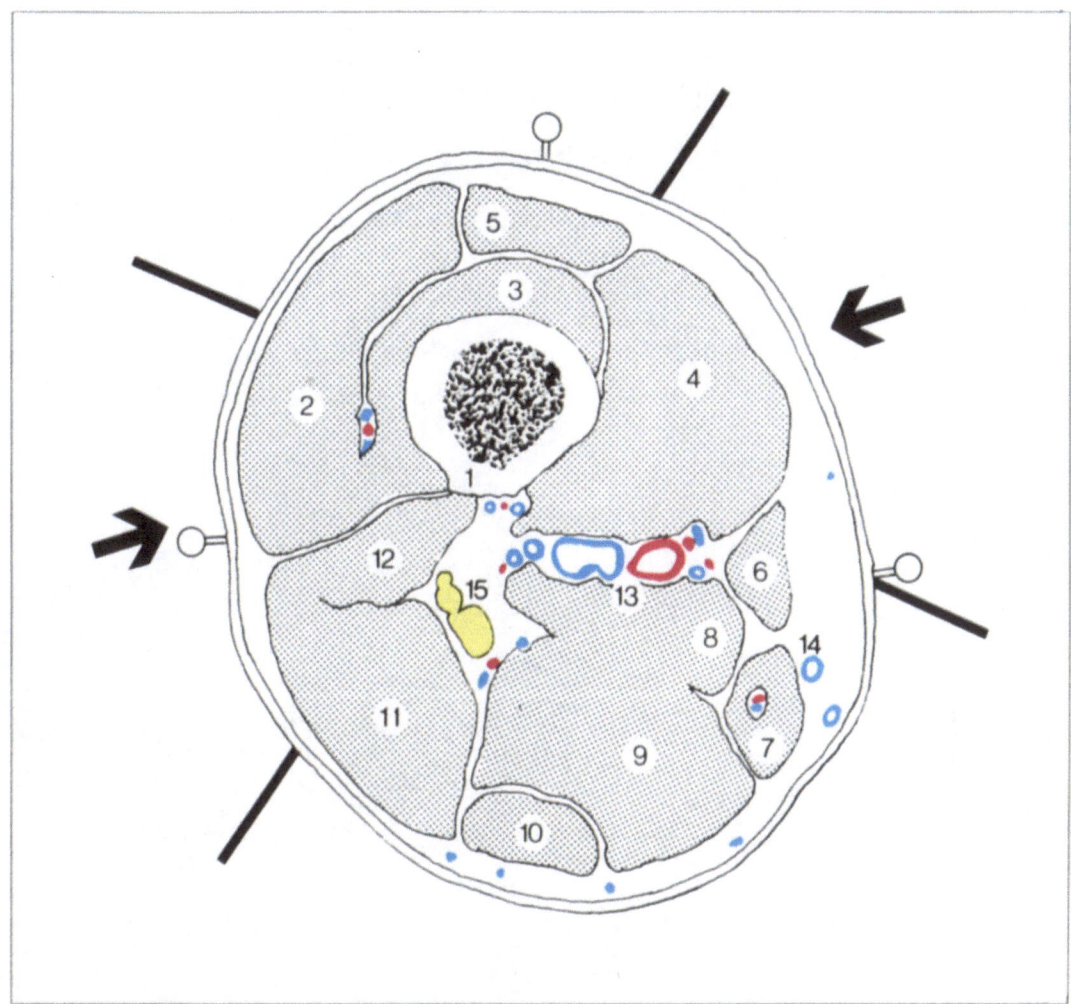

1 Femur	6 Sartorius	11 Biceps femoris (long head)
2 Vastus lateralis	7 Gracilis	12 Biceps femoris (short head)
3 Vastus intermedius	8 Adductor magnus	13 Femoral vessels
4 Vastus medialis	9 Semimembranosus	14 Long saphenous vein
5 Rectus femoris	10 Semitendinosus	15 Sciatic nerve

Safe Areas. These are large and lie laterally and medially in relation to the femur.

Cutaneous Zones Related to the Safe Areas. These comprise two zones lying laterally and ventro-medially in relation to the reference lines.

- *The ventro-medial zone* consists of the medial three-quarters of the area between the ventral and medial reference lines. The zone does not pass beyond the medial border of the rectus femoris muscle.

- *The lateral zone* extends symmetrically on either side of the axis of the lateral reference line and includes the area between the middle of the body of the vastus lateralis muscle to the middle of the body of the biceps femoris muscle.

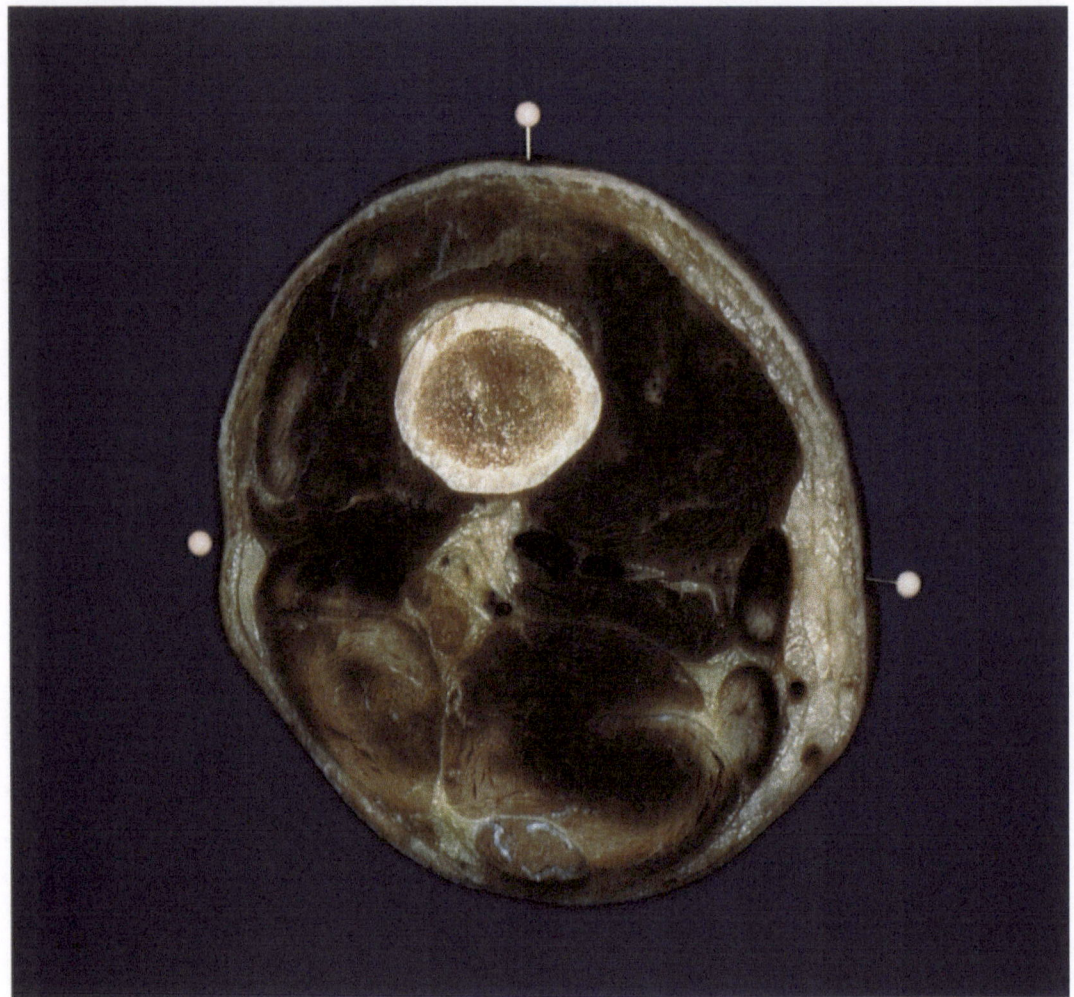

Cross-section C9

Bone. The section has been taken immediately distal to the junction of the middle and distal thirds of the diaphysis. The bone, which has increased in diameter, lies in the ventral half of the section.

Vessels and Nerves. The femoral vessels now approach the sciatic nerve, which has formed branches.

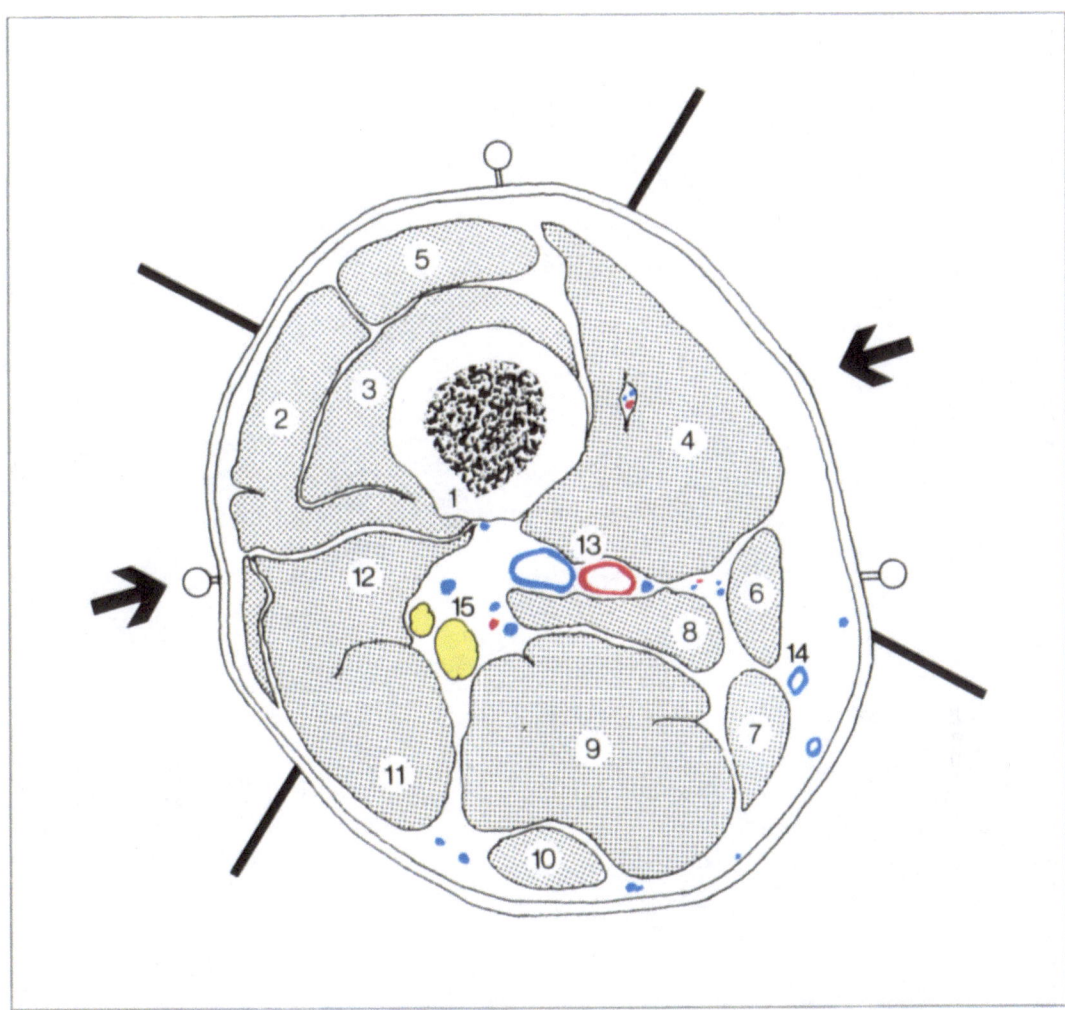

1 Femur	6 Sartorius	11 Biceps femoris (long head)
2 Vastus lateralis	7 Gracilis	12 Biceps femoris (short head)
3 Vastus intermedius	8 Adductor magnus	13 Femoral vessels
4 Vastus medialis	9 Semimembranosus	14 Long saphenous vein
5 Rectus femoris	10 Semitendinosus	15 Sciatic nerve

Safe Areas. These are large and lie laterally and medially in relation to the bone.

Cutaneous Zones Related to the Safe Areas. These comprise two zones lying laterally and ventro-medially in relation to the reference lines.

– *The ventro-medial zone* consists of the medial two-thirds of the area between the ventral and medial reference lines. This zone corresponds to the superficial portion of the vastus medialis muscle and is bounded medially by the sartorius muscle.

– *The lateral zone* extends symmetrically on either side of the axis of the lateral reference line. The zone extends from the middle of the body of the vastus lateralis to the middle of the body of the biceps femoris muscles.

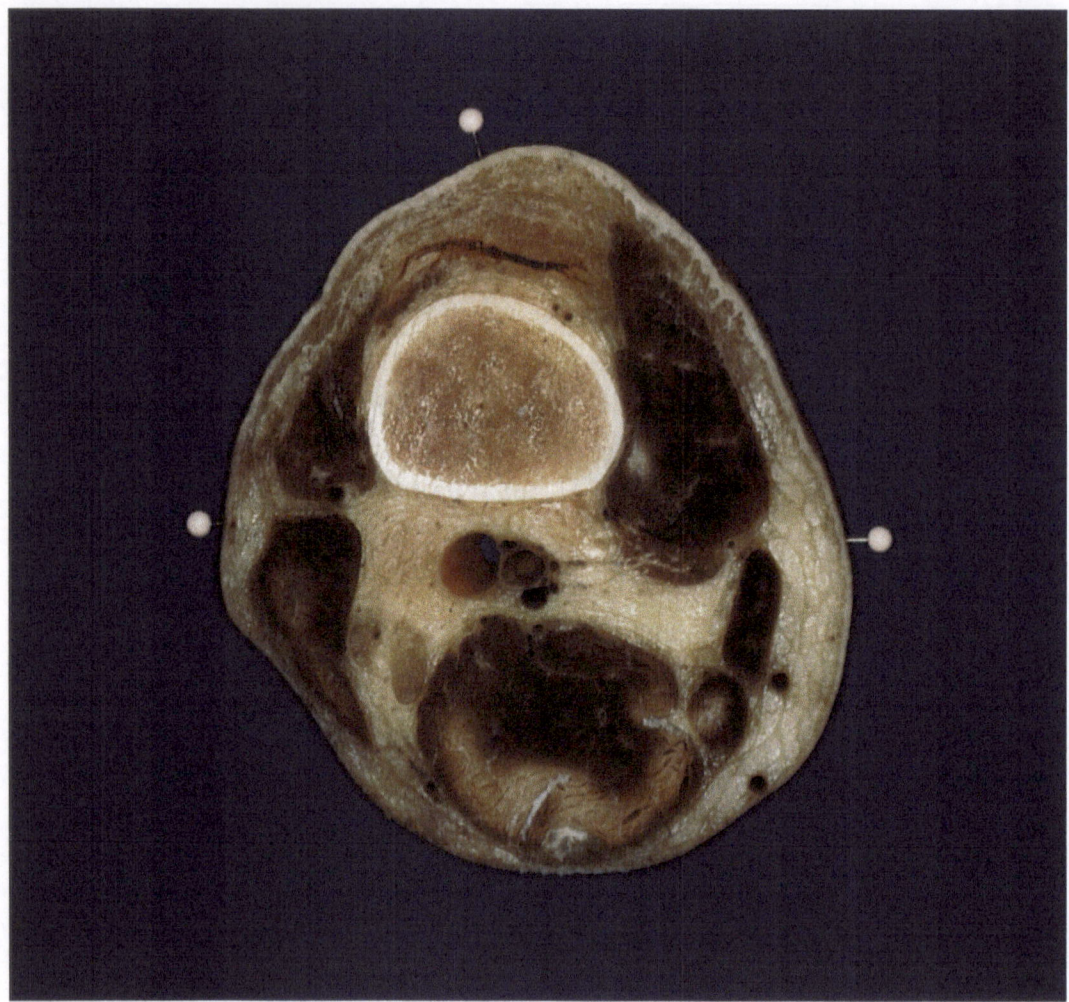

Cross-section C10

Bone. The section shows the femoral metaphysis. The diameter of the femur has become considerably larger.

Joint. The supra-patellar bursa shown in the section should not be transfixed.

Vessels and Nerves. The popliteal vessels are in contact with the dorsal aspect of the distal metaphysis. The sciatic nerve has divided into the common peroneal and tibial nerves, the latter of which lies more superficially and laterally.

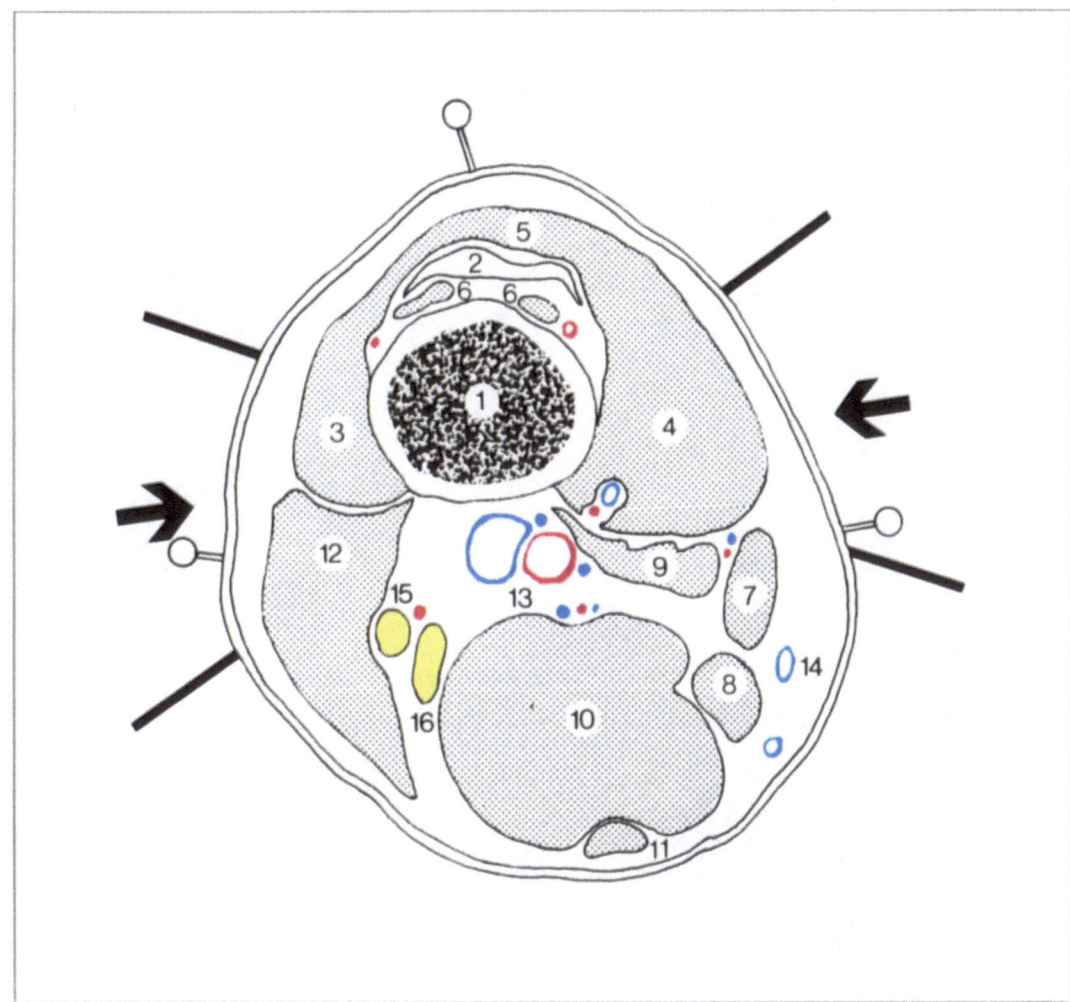

1 Femur (distal metaphysis)	7 Sartorius	12 Biceps femoris
2 Knee joint (supra-patellar bursa)	8 Gracilis	13 Popliteal vessels
3 Vastus lateralis	9 Adductor magnus	14 Long saphenous vein
4 Vastus medialis	10 Semimembranosus	15 Common peroneal nerve
5 Quadriceps (tendon)	11 Semitendinosus	16 Tibial nerve
6 Articularis genu		

Safe Areas. These have diminished as a result of the proximity of the vessels and the joint and are found directly medially and laterally in relation to the femur.

Cutaneous Zones Related to the Safe Areas. These comprise two zones lying opposite each other medially and laterally in relation to the reference lines.

– *The medial zone* includes the area from the medial reference line (i.e., the ventral aspect of the sartorius muscle) to the middle of the body of the vastus medialis muscle.

– *The lateral zone* is situated symmetrically about the medio-lateral axis and extends from the middle of the body of the biceps femoris muscle to the middle of the body of the vastus lateralis muscle.

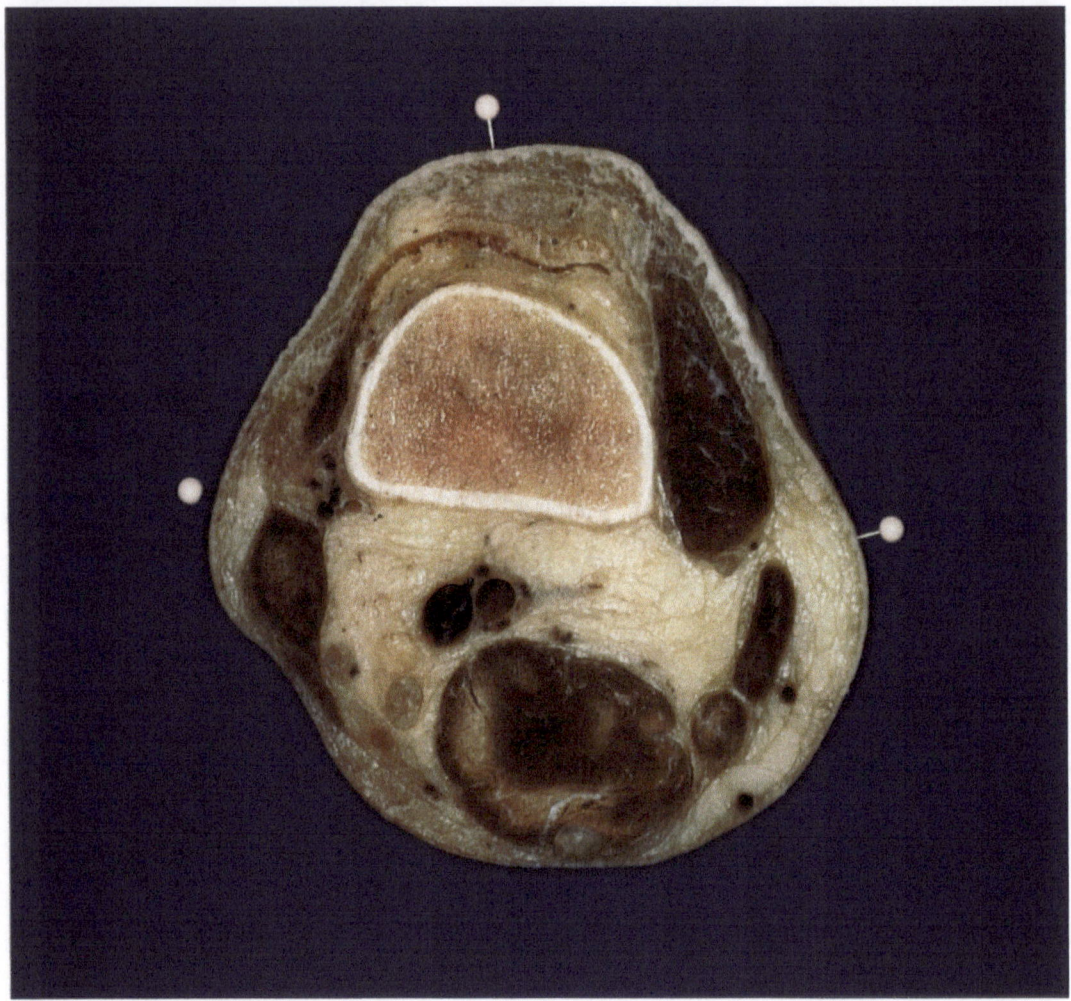

Cross-section C11

Bone. The section, taken immedately distal to Section C 10, shows a considerably expanded metaphysis.

Joint. The supra-patellar bursa shown in the section should not be transfixed.

Vessels and Nerves. The popliteal vessels and the tibial nerve are in close proximity to one another at this level and are located more superficially behind the expanded metaphysis. The common peroneal nerve lies directly against the dorso-medial border of the biceps femoris muscle.

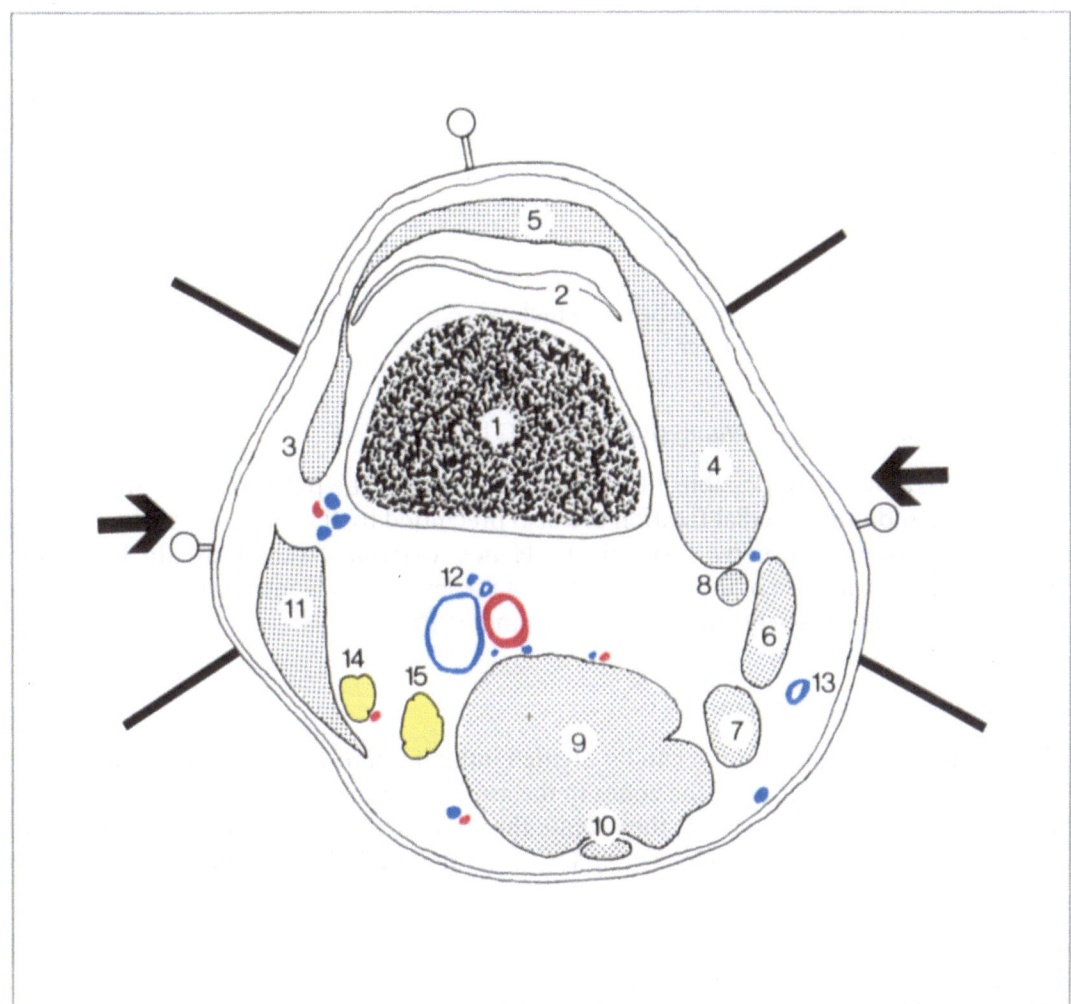

1 Femur (distal metaphysis)	6 Sartorius	11 Biceps femoris
2 Knee joint (supra-patellar bursa)	7 Gracilis	12 Popliteal vessels
3 Vastus lateralis	8 Adductor magnus	13 Long saphenous vein
4 Vastus medialis	9 Semimembranosus	14 Common peroneal nerve
5 Quadriceps (tendon)	10 Semitendinosus	15 Tibial nerve

Safe Areas. These are moderately large and are found directly laterally and medially in relation to the femur.

Cutaneous Zones Related to the Safe Areas. These comprise two zones lying medially and laterally in relation to the reference lines.

Their ventral limits correspond to the ventral borders of the bodies of the vasti medialis and lateralis muscles respectively, while dorsally, the two zones end slightly dorsal to the medial and lateral reference lines.

Safe Zones of the Thigh

The safe zones can be viewed as bands which wind around the thigh from its most proximal point to just above the knee.

Proximal Epiphysis and Metaphysis (Sections C1–C3)

External fixation is not possible either in the direction of the pelvis or away from it. Ventro-medial and dorso-medial fixation are also precluded by the presence of the profunda vessels and the sciatic nerve respectively. Hence, external fixation should be avoided through the prominence of the lateral aspect of the greater trochanter, but is permissible in the two areas immediately adjacent dorsally and ventrally.

Proximal Diaphysis (Sections C4, C5)

The circumference of the thigh can be divided into alternately safe and prohibited zones. These are the continuation of the zones in the more proximal sections but as they proceed towards the level of Section C5, they rotate through 30°, running parallel to the femoral artery and the sartorius muscle. The latter forms the medial limit, which should not be breached.

Middle and Distal Diaphysis (Sections C6–C9)

The four bands continue to rotate as they descend towards the knee. At the level of Section C9, the bands have rotated through a further 30°. The sartorius muscle remains the medial limit beyond which external fixation should not be attempted.

Distal Metaphysis (Sections C10, C11)

The four bands continue to rotate in the same direction as the sartorius muscle, and by the level of Section C11, a further rotation of 30° has occurred.
Thus, between the trochanteric and supra-patellar regions, the safe zones rotate through 90°. In the most proximal region, they lie strictly ventrally and dorsally, progressively winding in a distal direction until they lie strictly medially and laterally at the level of the knee. They follow the same course as the sartorius muscle which always represents the medial limit for external fixation.

The safe zone for external fixation that initially lies dorsally has been called the dorso-lateral safe zone (band 1) whereas the one that is initially located ventrally has been named the ventro-medial safe zone (band 2).

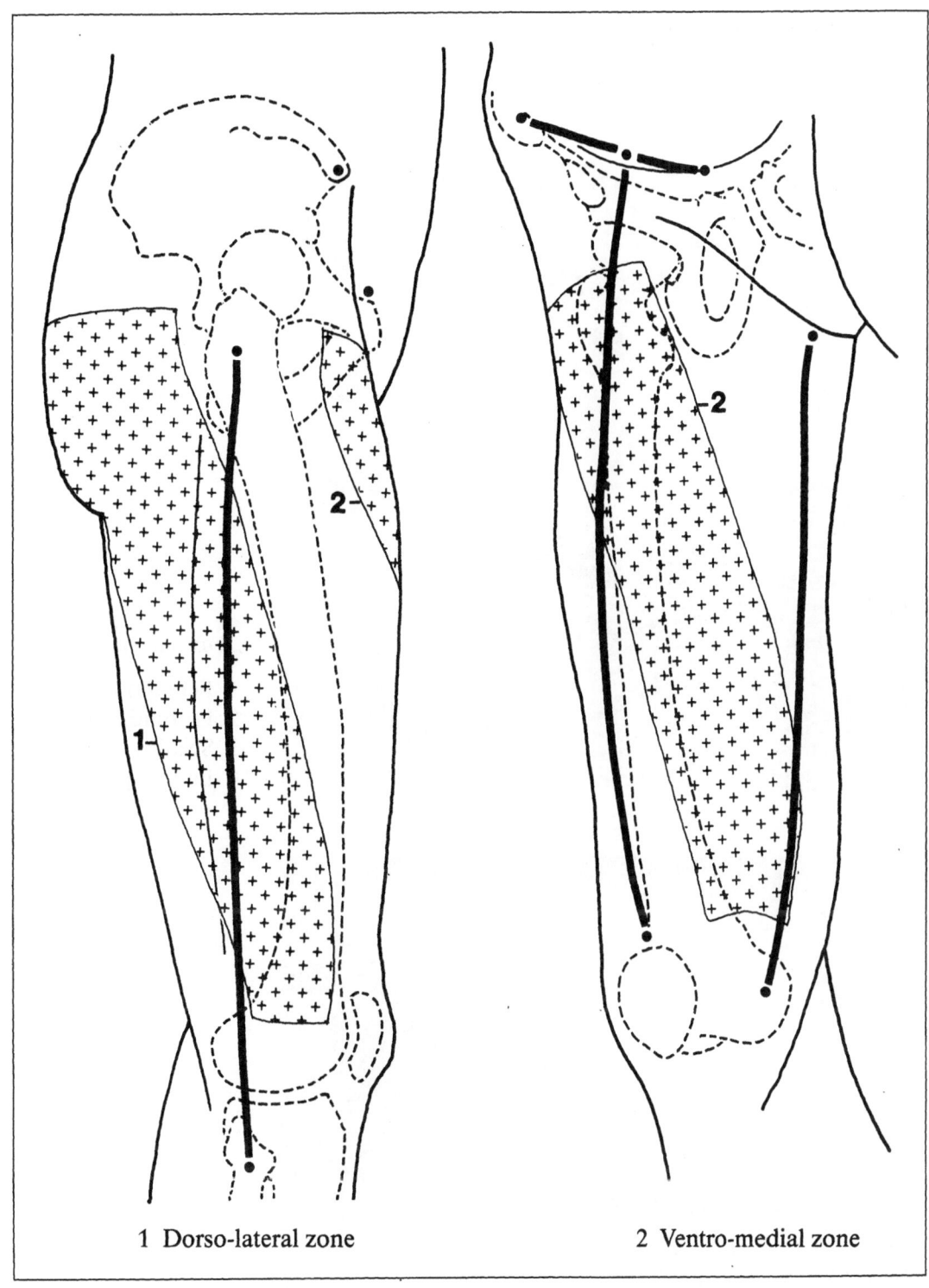

1 Dorso-lateral zone   2 Ventro-medial zone

Computerised Tomographic Scans of the Thigh

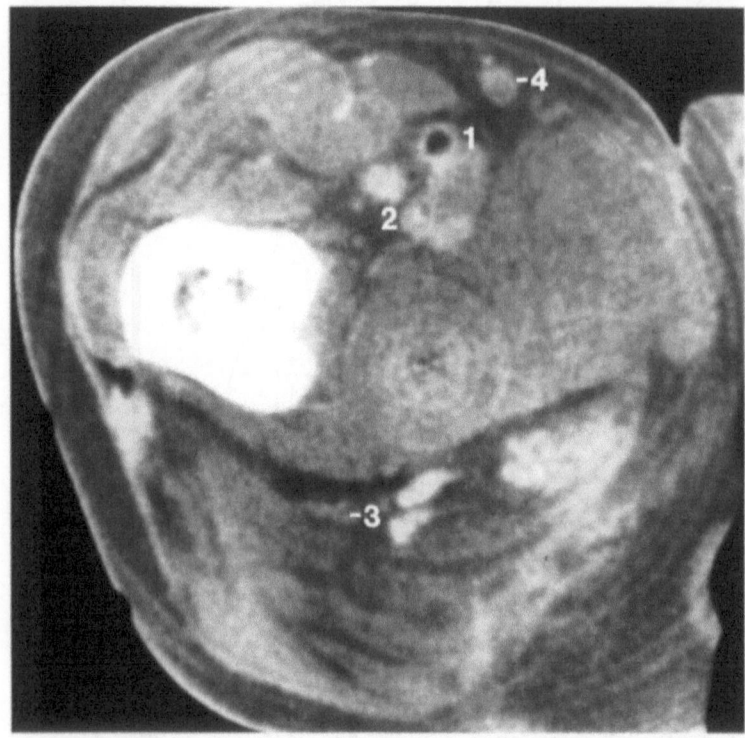

Proximal epiphysis

1 Femoral vessels
2 Profunda vessels
3 Sciatic nerve
4 Long saphenous vein

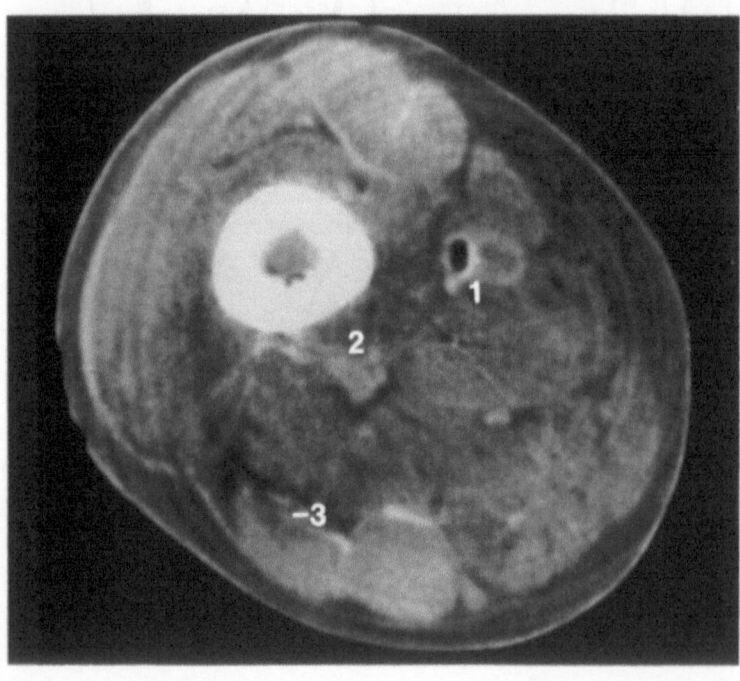

Proximal diaphysis

1 Femoral vessels
2 Profunda vessels
3 Sciatic nerve

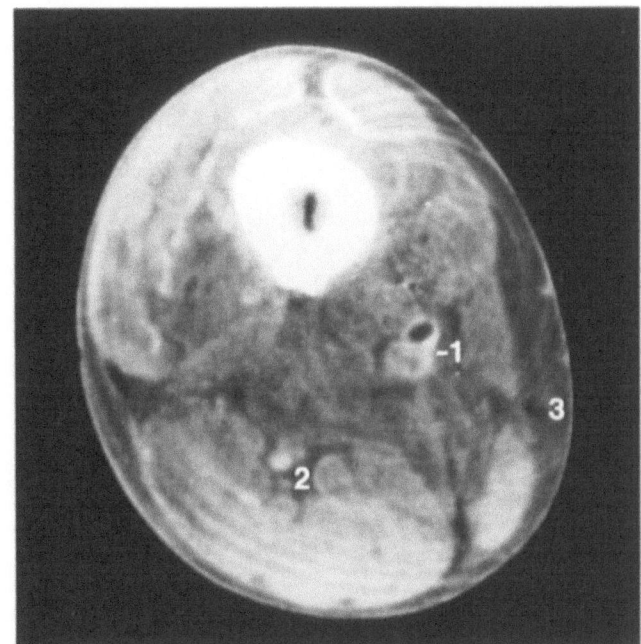

Distal diaphysis

1 Femoral vessels
2 Sciatic nerve
3 Long saphenous vein

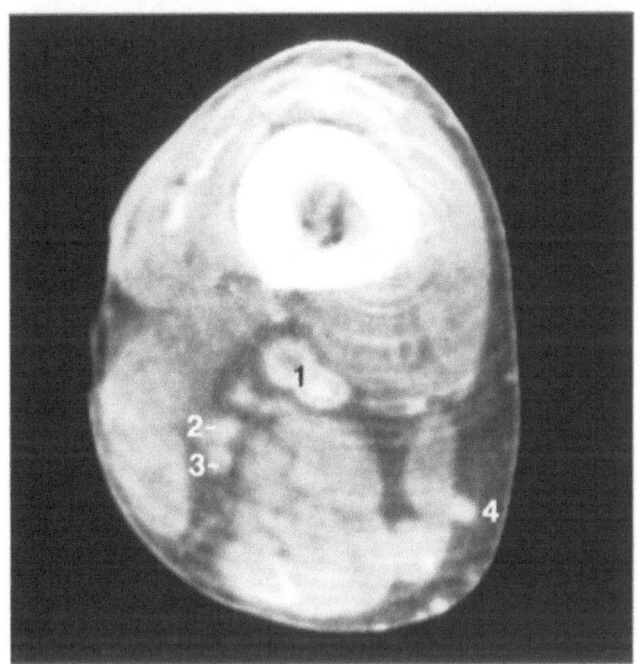

Distal metaphysis

1 Popliteal vessels
2 Common peroneal nerve
3 Tibial nerve
4 Long saphenous vein

D. Cross-sections of the Leg

Levels of the Cross-sections

The 11 cross-sections are arranged into five groups corresponding to areas commonly used for external fixation.

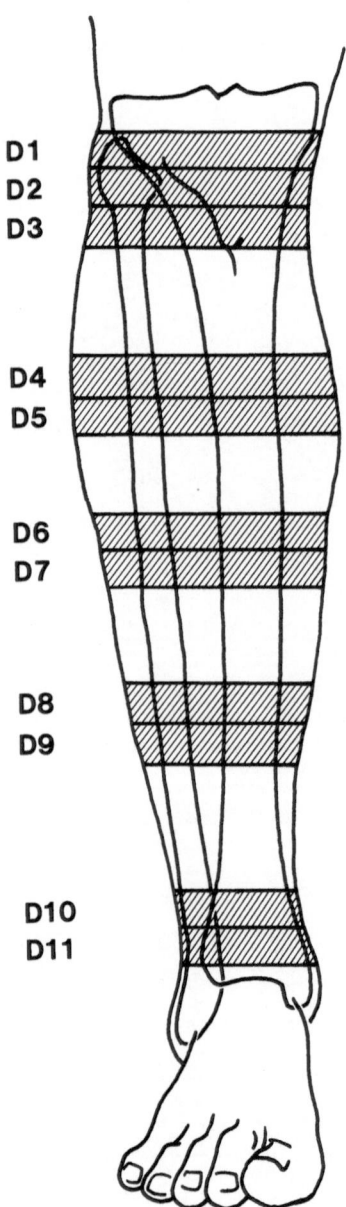

Proximal Epiphysis and Metaphysis (D1-D3)

The sections have been taken between the proximal tibio-fibular joint and the tibial tuberosity.

Proximal Diaphysis (D4, D5)

The sections have been taken on either side of the junction of the proximal and middle thirds of the diaphysis.

Middle Diaphysis (D6, D7)

The sections have been taken through the midpoint of the diaphysis.

Distal Diaphysis (D8, D9)

The sections have been taken on either side of the junction of the middle and distal thirds of the diaphysis.

Distal Metaphysis and Epiphysis (D10, D11)

The sections have been taken adjacent to the malleoli.

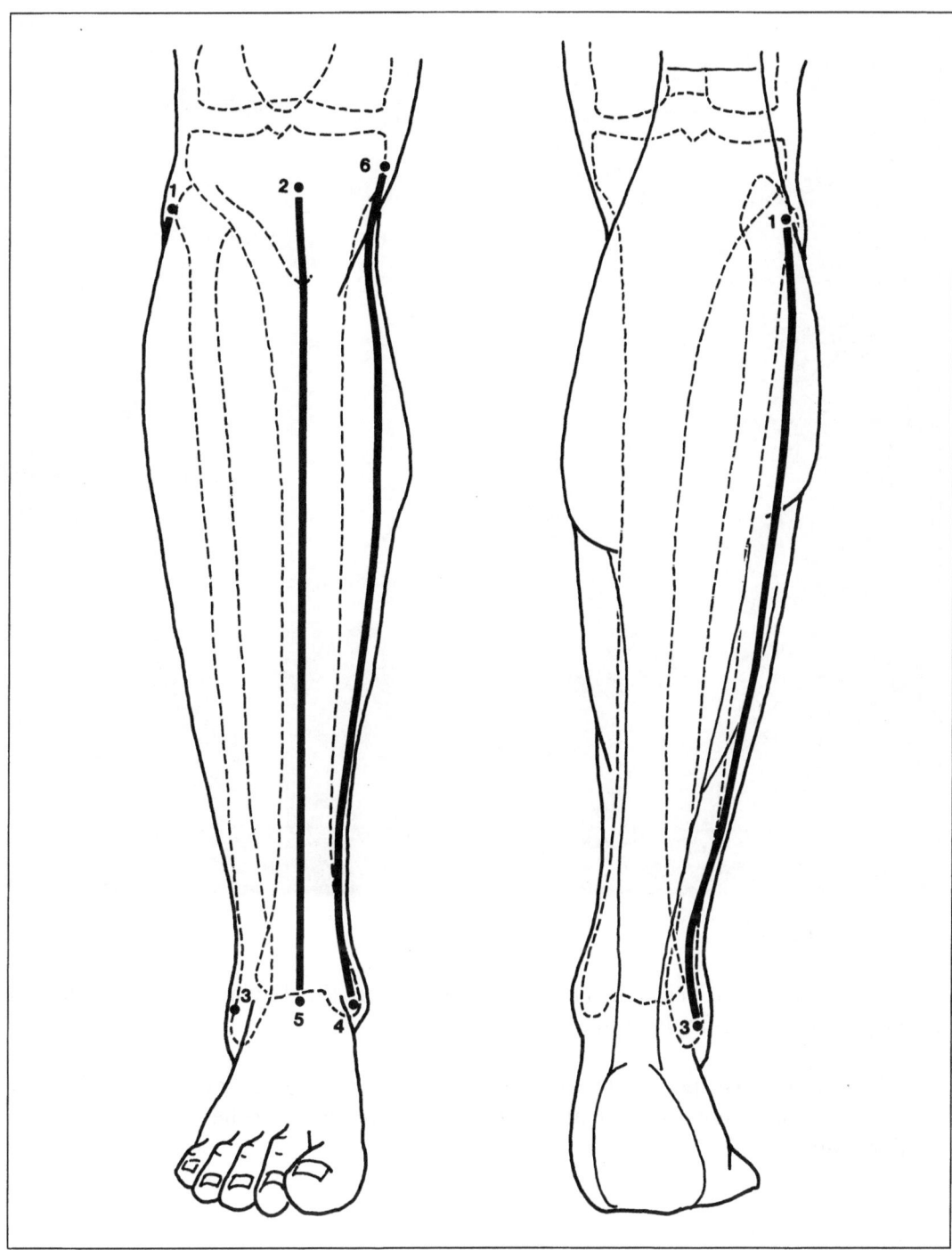

Landmarks

1 Middle of the fibular head
2 Tibial tuberosity
3 Middle of the lateral malleolus
4 Distal end of the medial malleolus
5 Midpoint of line 3–4
6 Point diametrically opposite the fibular head

Reference lines

Ventral line: line 2–5
Lateral line: line 1–3
Medial line: line 4–6

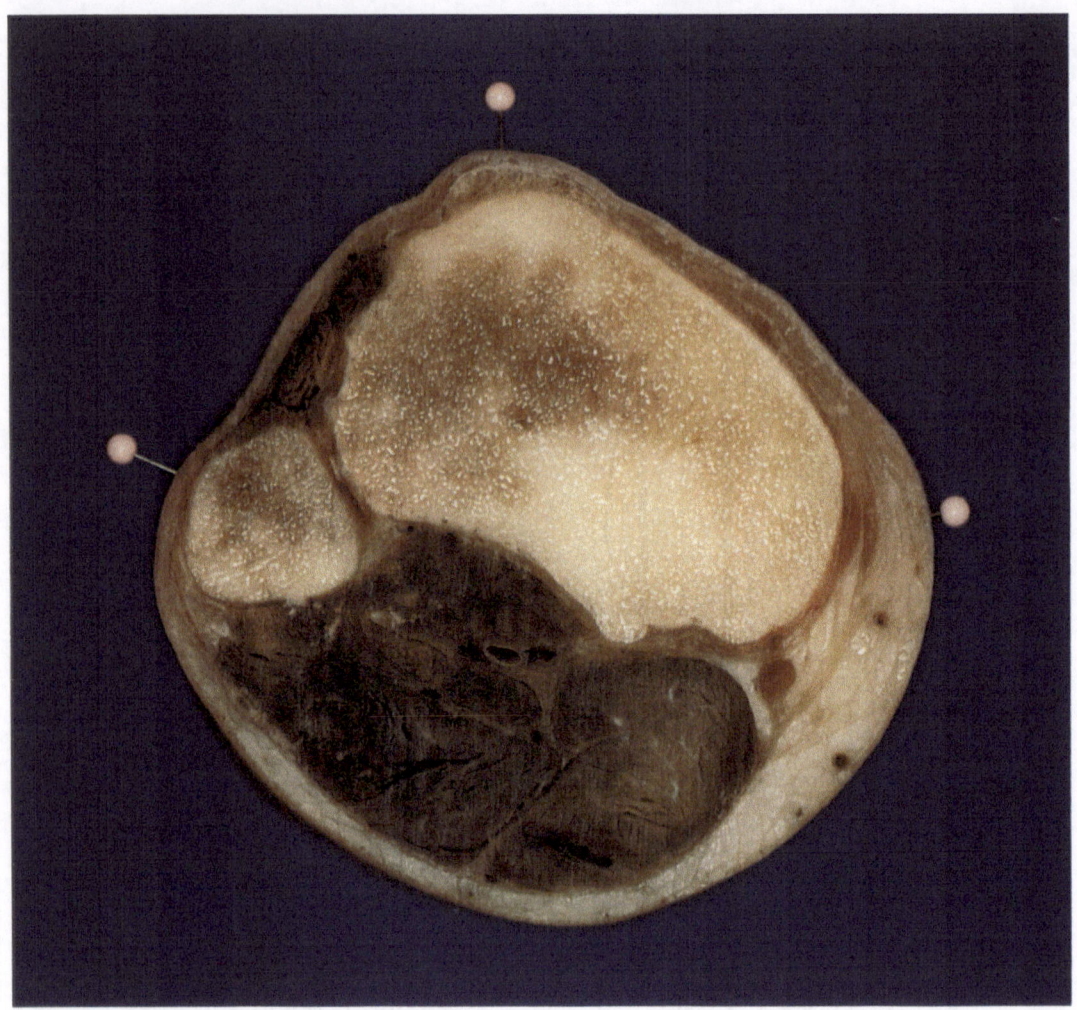

Cross-section D 1

Bones. The section has been taken through the broad proximal tibial epiphysis at the level of the proximal tibio-fibular joint.

Vessels and Nerves. The popliteal vessels lie in the median plane and deep. The tibial nerve lies close by, but is more superficial. The common peroneal nerve lies in contact with the fibular head.

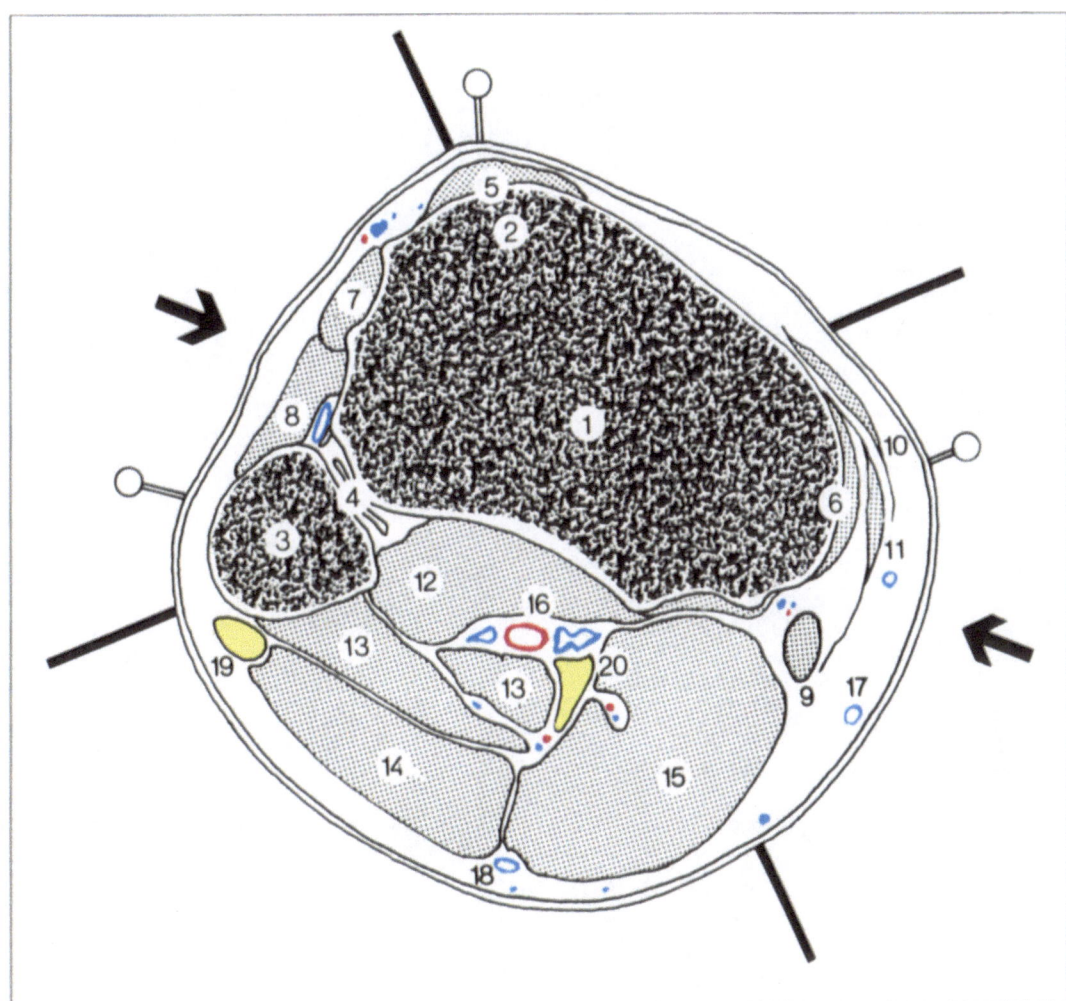

1 Tibia	11 Semitendinosus
2 Tibial tuberosity	12 Popliteus
3 Fibula	13 Soleus
4 Proximal tibio-fibular joint	14 Gastrocnemius (lateral head)
5 Patellar ligament	15 Gastrocnemius (medial head)
6 Tibial collateral ligament of the knee	16 Popliteal vessels
7 Tibialis anterior	17 Long saphenous vein
8 Extensor digitorum longus	18 Short saphenous vein
9 Gracilis	19 Common peroneal nerve
10 Sartorius	20 Tibial nerve

Safe Areas. These are large and lie medially and laterally in relation to the tibia.

Cutaneous Zones Related to the Safe Areas. These comprise two zones lying ventro-laterally and medially in relation to the reference lines.

- *The ventro-lateral zone* lies between the ventral and lateral reference lines (middle of the tibial tuberosity and the fibular head).

- *The medial zone* extends from the middle of the space between the ventral and medial reference lines to the medial border of the medial head of the gastrocnemius muscle.

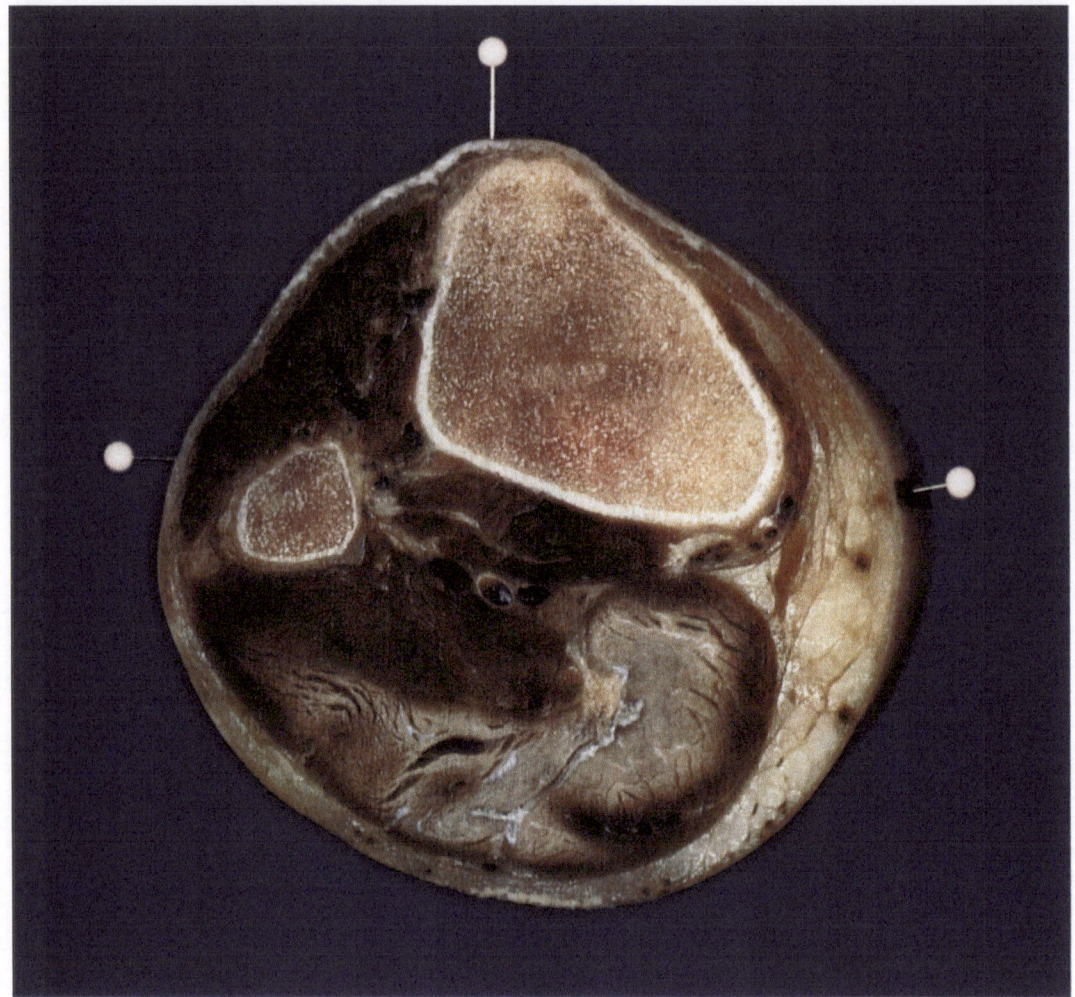

Cross-section D 2

Bones. The section shows the tibial metaphysis immediately distal to the proximal tibio-fibular joint.

Vessels and Nerves. The popliteal vessels and the tibial nerve lie in the median plane and deep. The common peroneal nerve passes ventrally along the neck of the fibula. Branches of the anterior tibial vessels are found between the tibialis anterior and extensor digitorum longus muscles.

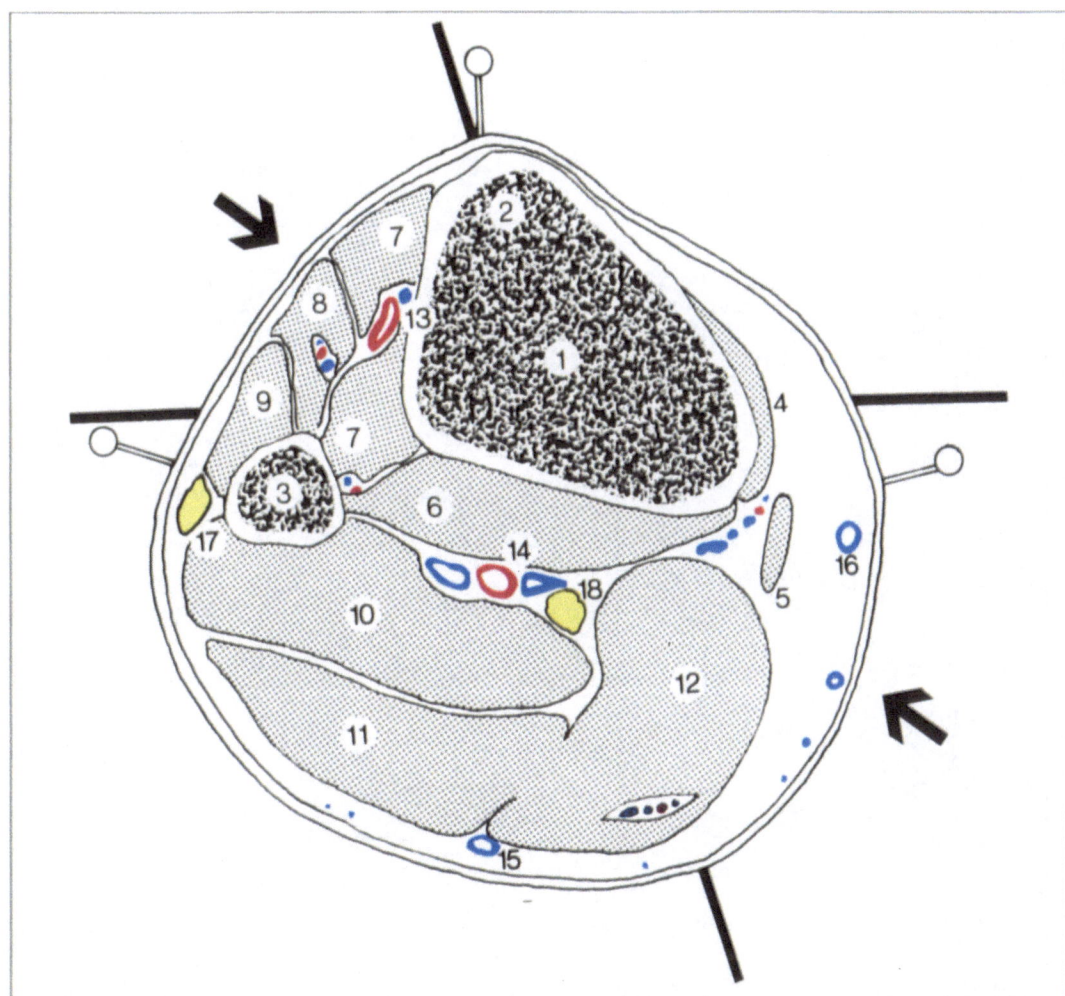

1 Tibia
2 Tibial tuberosity
3 Fibula (neck)
4 Tibial collateral ligament of the knee +
   semitendinosus + gracilis
5 Sartorius
6 Popliteus
7 Tibialis anterior
8 Extensor digitorum longus
9 Peroneus longus

10 Soleus
11 Gastrocnemius (lateral head)
12 Gastrocnemius (medial head)
13 Anterior tibial vessels (branches)
14 Posterior tibial vessels
15 Short saphenous vein
16 Long saphenous vein
17 Common peroneal nerve
18 Tibial nerve

Safe Areas. These are large and lie ventro-laterally and dorso-medially in relation to the tibia.

Cutaneous Zones Related to the Safe Areas. These consist of two zones lying ventro-laterally and dorso-medially in relation to the reference lines.

- *The ventro-lateral zone* lies between the ventral and lateral reference lines, ending just ventral to the fibular neck, upon which lies the common peroneal nerve.

- *The dorso-medial zone* lies between the medial border of the tibia and the middle of the body of the medial head of the gastrocnemius muscle.

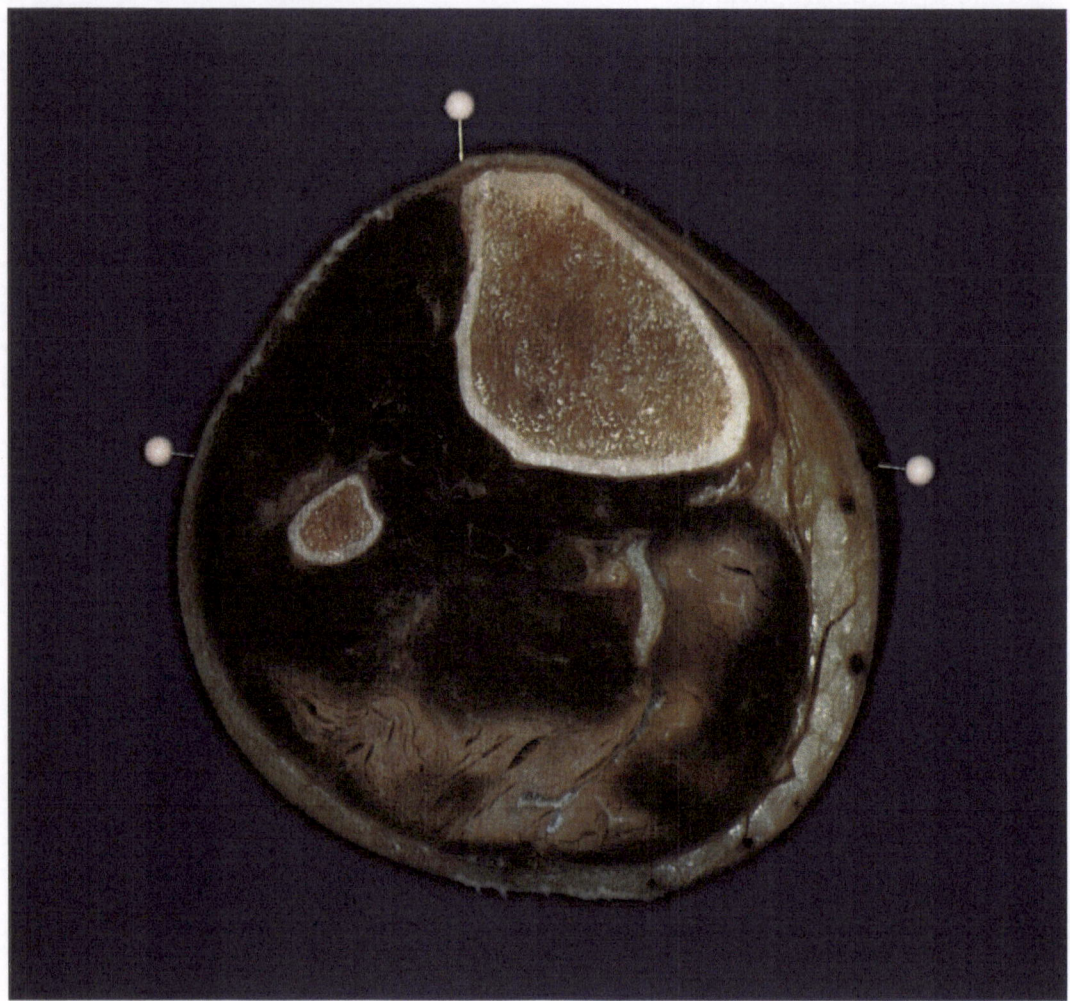

Cross-section D 3

Bones. The section has been taken through the tibial tuberosity. The width of the tibia has diminished, and the bone now lies in the ventro-medial quadrant.

Vessels and Nerves. The popliteal vessels divide into anterior and posterior tibial vessels. The tibial nerve lies medial to the vessels in the same transverse plane as the fibula. The common peroneal nerve divides into the deep peroneal nerve (ventral) and the superficial peroneal nerve (lateral).

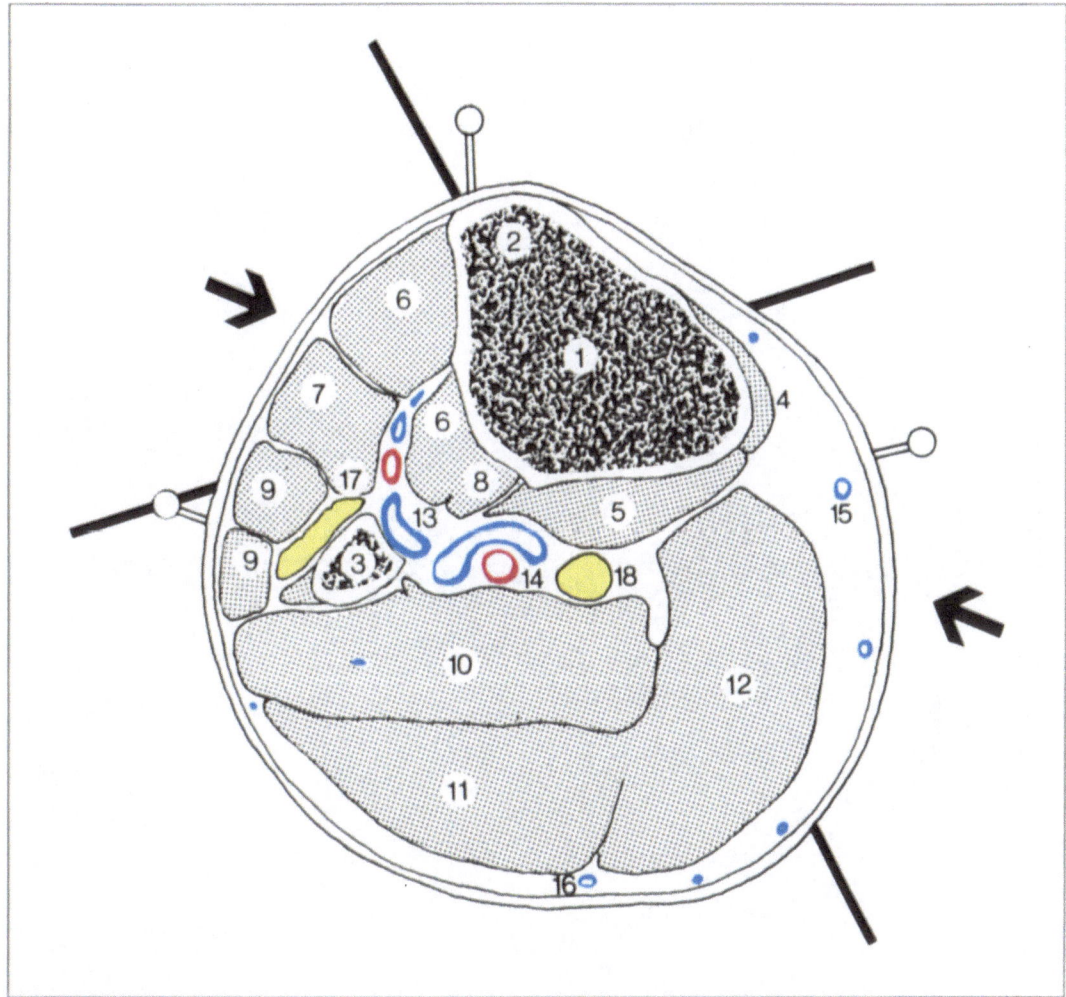

1 Tibia	10 Soleus
2 Tibial tuberosity	11 Gastrocnemius (lateral head)
3 Fibula	12 Gastrocnemius (medial head)
4 Tibial collateral ligament of the knee + sartorius + gracilis	13 Anterior tibial vessels
5 Popliteus	14 Posterior tibial vessels
6 Tibialis anterior	15 Long saphenous vein
7 Extensor digitorum longus	16 Short saphenous vein
8 Tibialis posterior	17 Common peroneal nerve
9 Peroneus longus	18 Tibial nerve

Safe Areas. These are large and lie ventro-laterally and medially in relation to the tibia.

Cutaneous Zones Related to the Safe Areas. These comprise two zones lying ventro-laterally and dorso-medially in relation to the reference lines.

- *The ventro-lateral zone* lies between the ventral and lateral reference lines.

- *The larger dorso-medial zone* extends from the medial surface of the tibia to the middle of the body of the medial head of the gastrocnemius muscle.

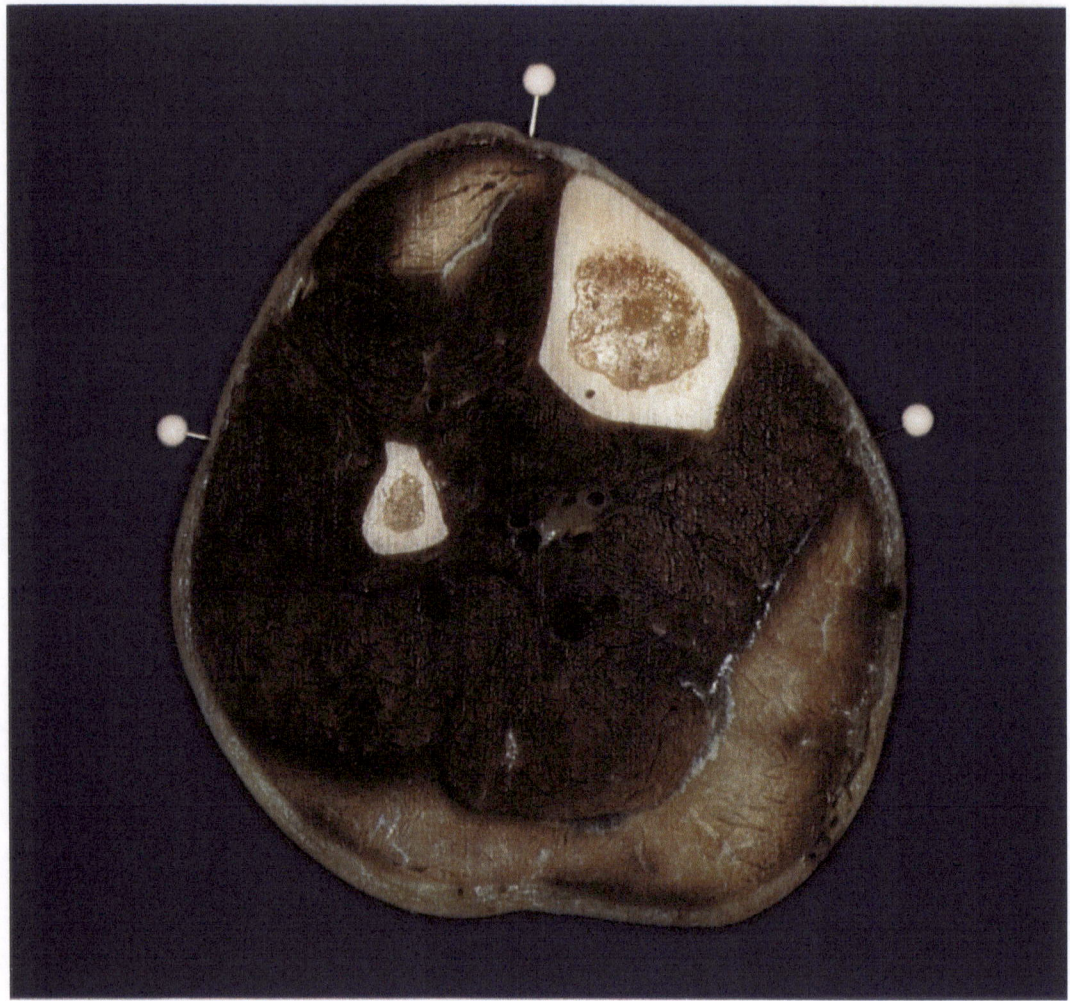

Cross-section D 4

Bones. The section has been taken proximal to the junction of the proximal and middle thirds of the tibia. The fibula lies 45° dorso-laterally in relation to the tibia.

Vessels and Nerves. The anterior tibial vessels and the deep peroneal nerve lie close to the ventral crest of the fibula. The superficial peroneal nerve lies under the peroneus longus muscle. The peroneal vessels diverge from the posterior tibial vessels, which accompany the tibial nerve.

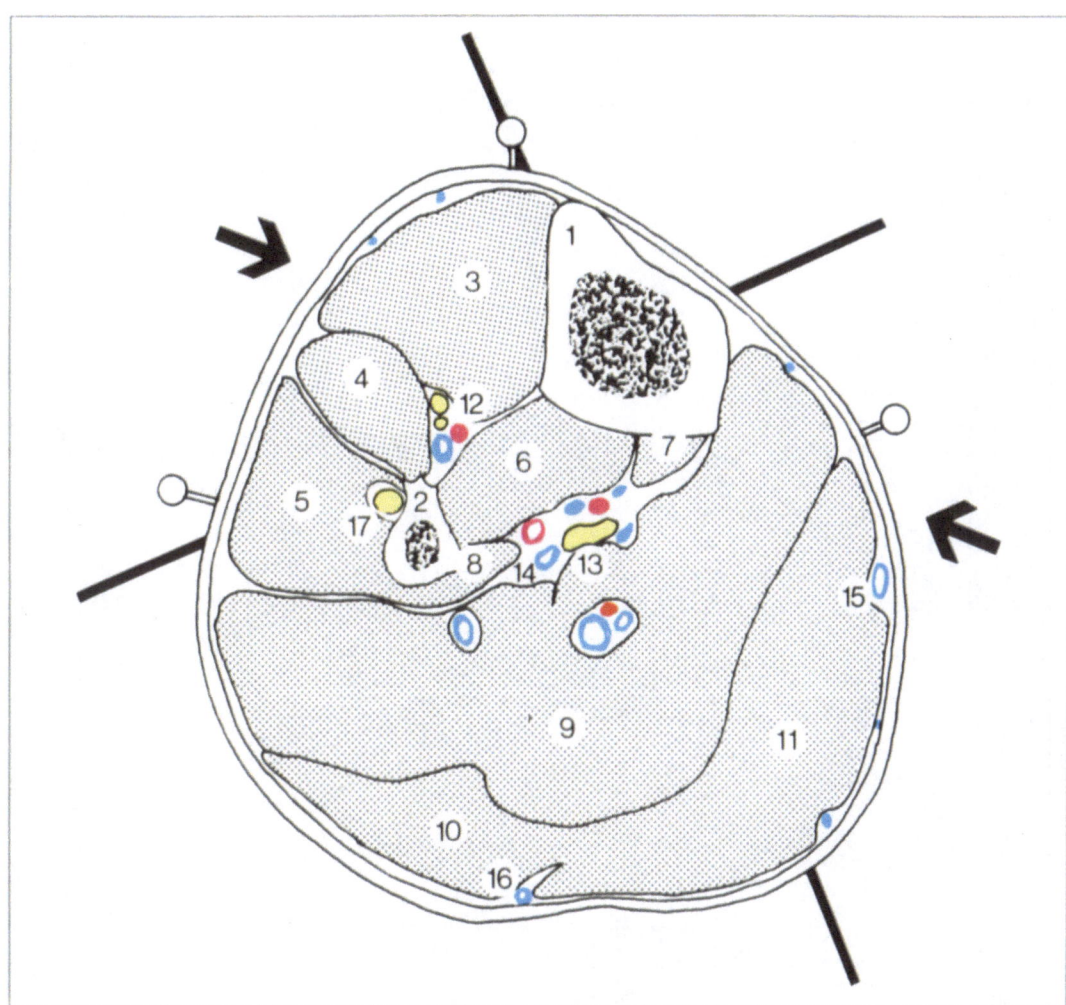

1 Tibia
2 Fibula
3 Tibialis anterior
4 Extensor digitorum longus
5 Peroneus longus
6 Tibialis posterior
7 Flexor digitorum longus
8 Flexor hallucis longus
9 Soleus

10 Gastrocnemius (lateral head)
11 Gastrocnemius (medial head)
12 Anterior tibial vessels + deep peroneal nerve
13 Posterior tibial vessels + tibial nerve
14 Peroneal vessels
15 Long saphenous vein
16 Short saphenous vein
17 Superficial peroneal nerve

Safe Areas. These are large and lie ventro-laterally and dorso-medially in relation to the tibia.

Cutaneous Zones Related to the Safe Areas. These comprise two zones lying ventro-laterally and dorso-medially in relation to the reference lines.

- *The ventro-lateral zone* lies between the ventral and lateral reference lines.

- *The dorso-medial zone* lies between the medial border of the tibia and the middle of the body of the medial head of the gastrocnemius muscle.

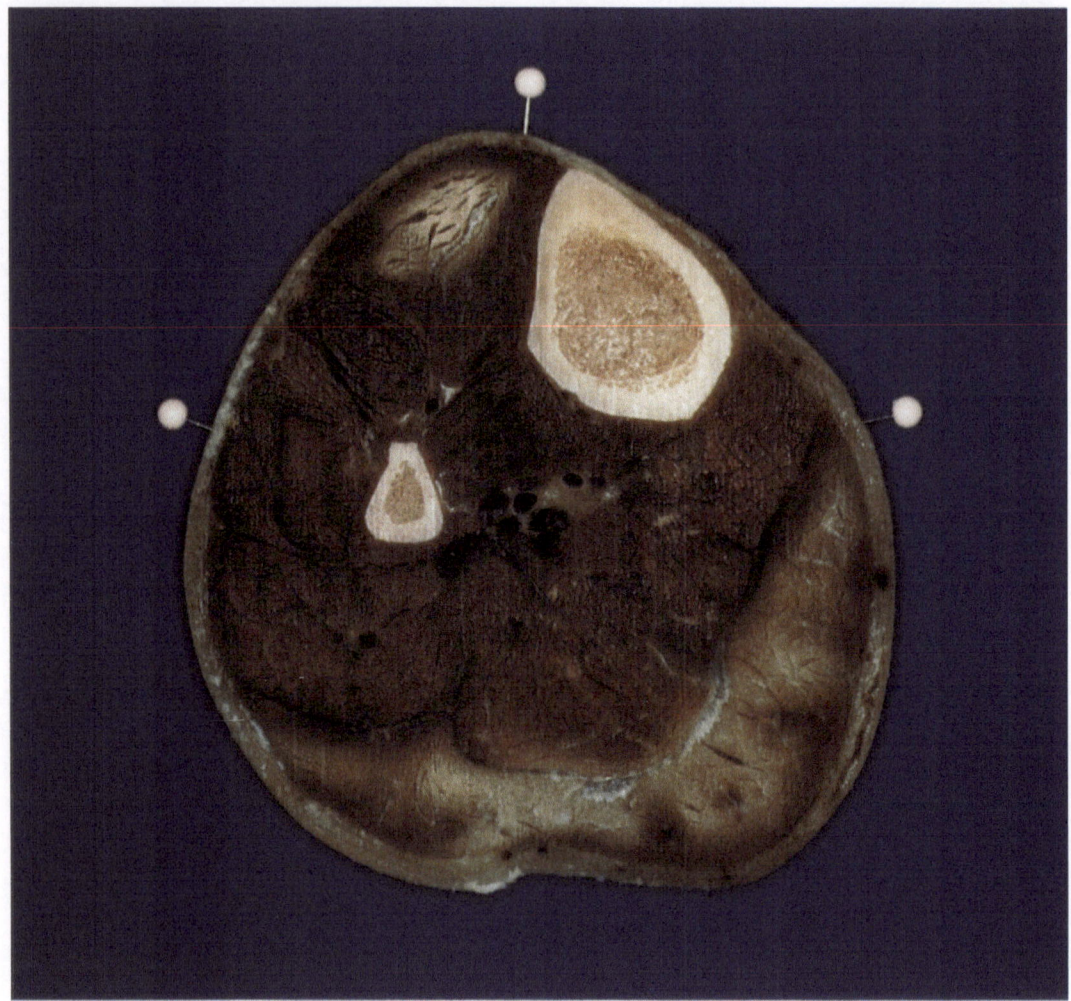

Cross-section D 5

Bones. The section has been taken distal to the junction of the proximal and middle thirds of the tibia.

Vessels and Nerves. The anterior tibial vessels and the deep peroneal nerve lie ventral to the ventral crest of the fibula. The superficial peroneal nerve lies on the lateral aspect of the fibula. The tibial nerve and the posterior tibial vessels lie dorsal to the tibia; the peroneal vessels are separate from them.

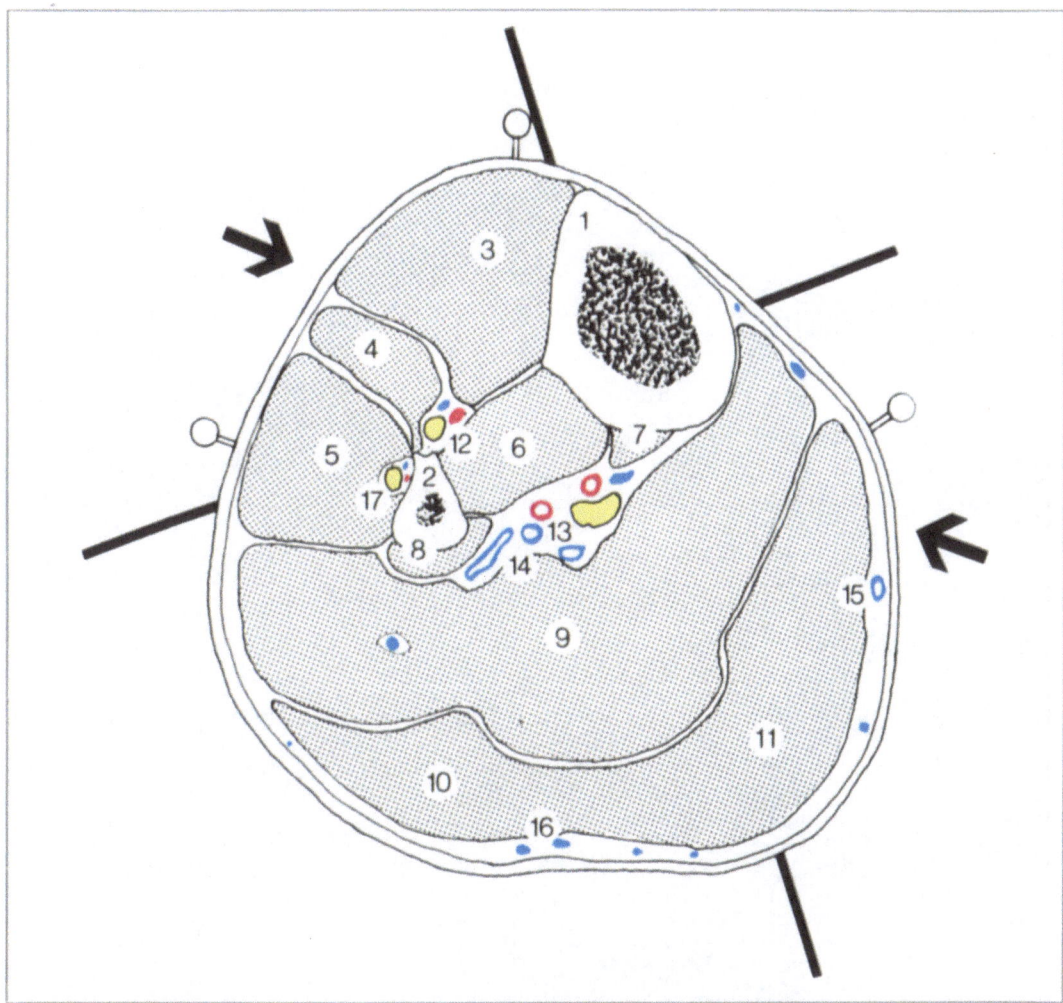

1 Tibia	10 Gastrocnemius (lateral head)
2 Fibula	11 Gastrocnemius (medial head)
3 Tibialis anterior	12 Anterior tibial vessels + deep peroneal nerve
4 Extensor digitorum longus	13 Posterior tibial vessels + tibial nerve
5 Peroneus longus	14 Peroneal vessels
6 Tibialis posterior	15 Long saphenous vein
7 Flexor digitorum longus	16 Short saphenous vein
8 Flexor hallucis longus	17 Superficial peroneal nerve
9 Soleus	

Safe Areas. These are large and lie ventro-laterally and medially in relation to the tibia.

Cutaneous Zones Related to the Safe Areas. These comprise two zones lying ventro-laterally and dorso-medially in relation to the reference lines.

- *The ventro-lateral zone* lies between the ventral and lateral reference lines.

- *The dorso-medial zone* lies between the middle of the medial aspect of the tibia and the middle of the body of the medial head of the gastrocnemius muscle.

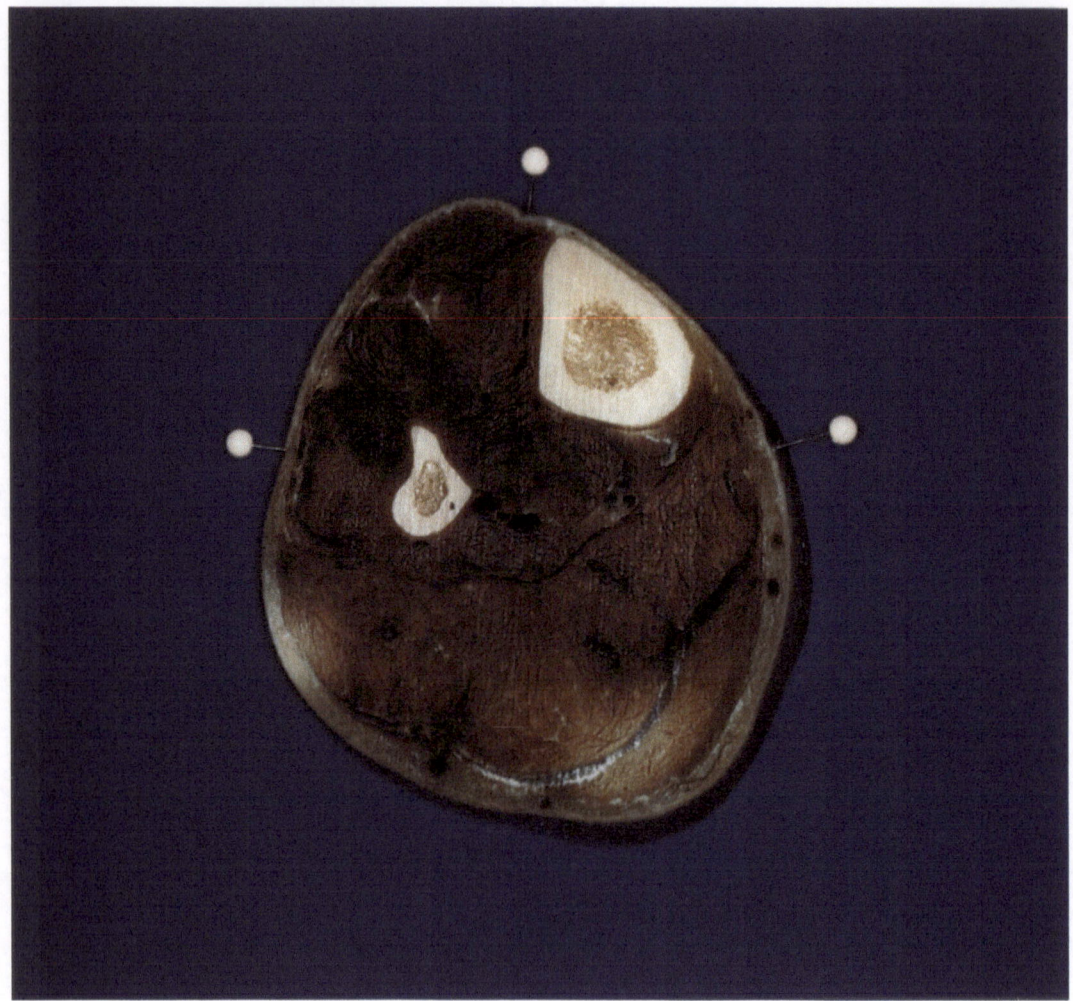

Cross-section D 6

Bones. The section has been taken just proximal to the middle of the tibial diaphysis.

Vessels and Nerves. The anterior tibial vessels and the deep peroneal nerve are found in the middle of the interosseous area. The posterior tibial vessels and the tibial nerve lie dorsal to the tibia. The superficial peroneal nerve lies between the peroneus longus and brevis muscles.

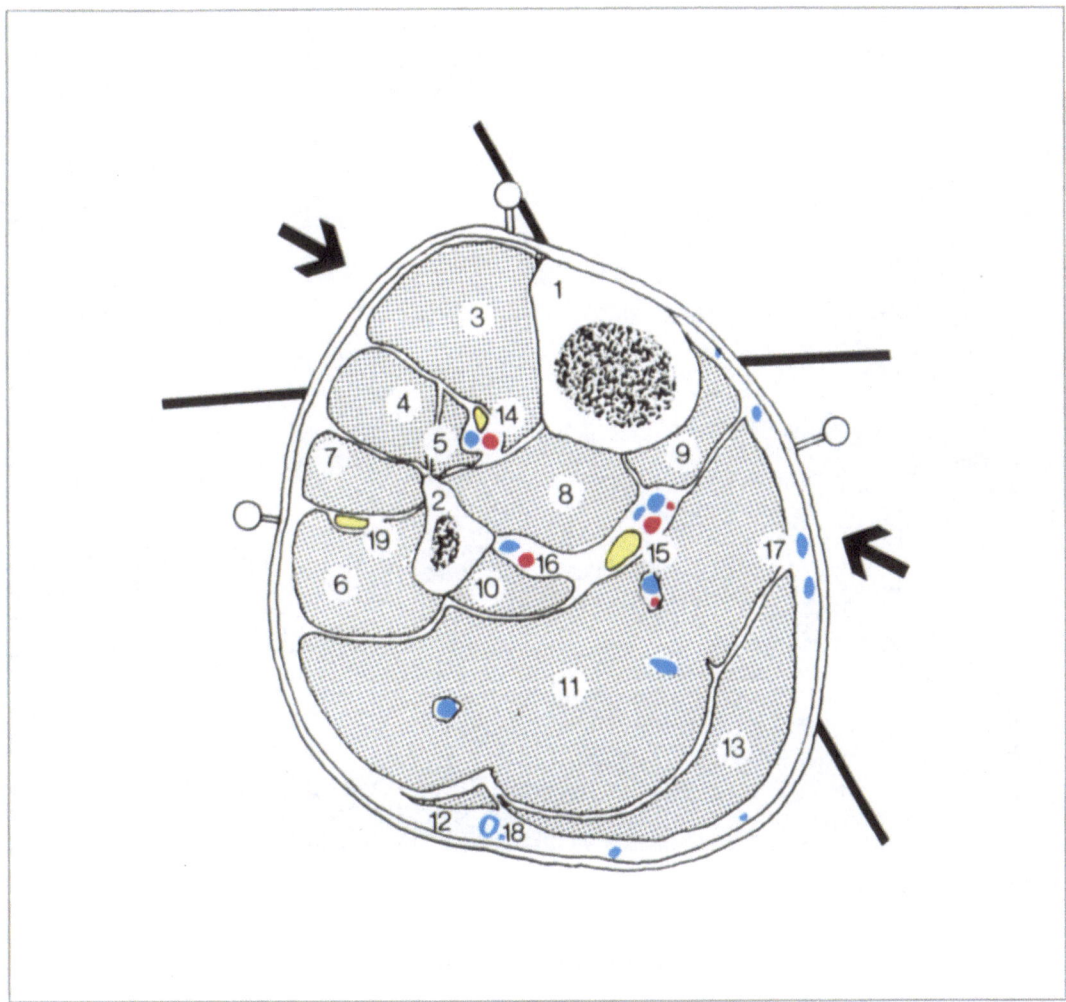

1 Tibia
2 Fibula
3 Tibialis anterior
4 Extensor digitorum longus
5 Extensor hallucis longus
6 Peroneus longus
7 Peroneus brevis
8 Tibialis posterior
9 Flexor digitorum longus
10 Flexor hallucis longus

11 Soleus
12 Gastrocnemius (lateral head)
13 Gastrocnemius (medial head)
14 Anterior tibial vessels + deep peroneal nerve
15 Posterior tibial vessels + tibial nerve
16 Peroneal vessels
17 Long saphenous vein
18 Short saphenous vein
19 Superficial peroneal nerve

Safe Areas. These are reduced and lie ventro-laterally and dorso-medially in relation to the tibia.

Cutaneous Zones Related to the Safe Areas. These comprise two zones lying ventro-laterally and dorso-medially in relation to the reference lines.

- *The ventro-lateral zone* consists of the ventral two-thirds of the area between the ventral and lateral reference lines.

- *The dorso-medial zone* lies between the middle of the medial aspect of the tibia and the medial border of the medial head of the gastrocnemius muscle.

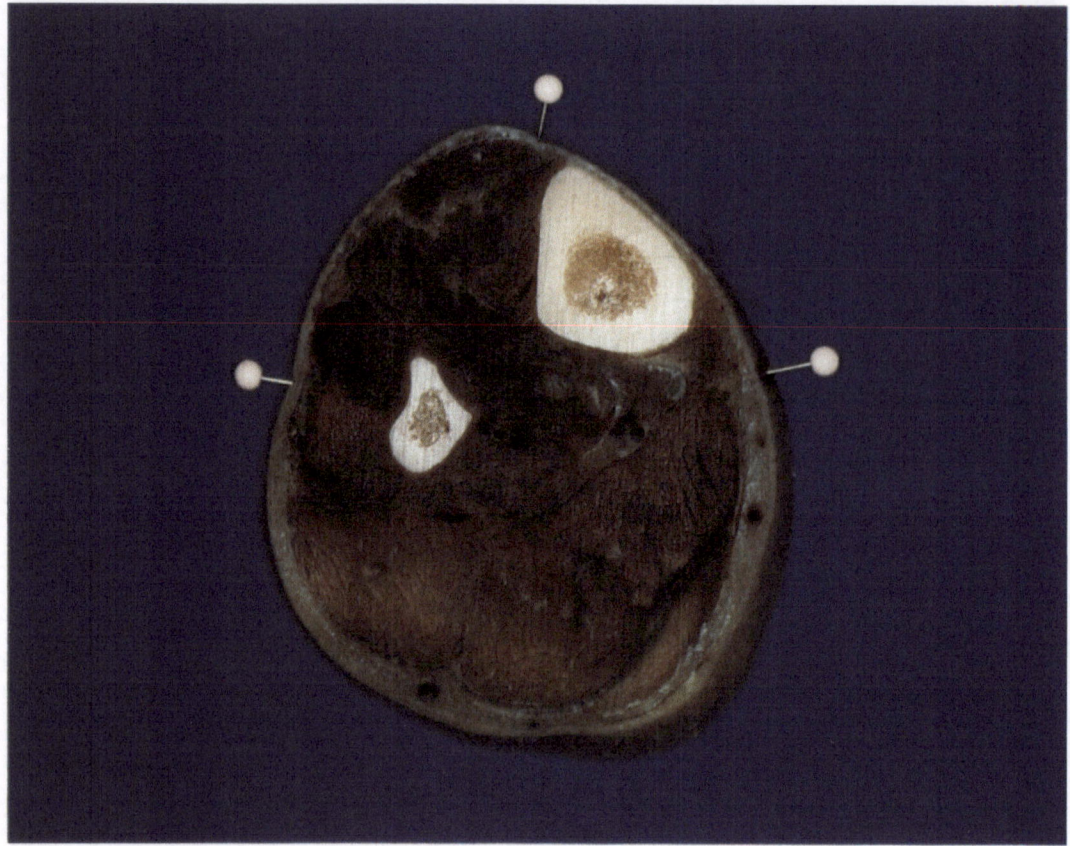

Cross-section D 7

Bones. The section has been taken just distal to the middle of the tibial diaphysis.

Vessels and Nerves. The anterior tibial vessels and the deep peroneal nerve are found in the middle of the interosseous area. The posterior tibial vessels and the tibial nerve lie dorsal to the tibia. The superficial peroneal nerve lies between the peroneus longus and brevis muscles.

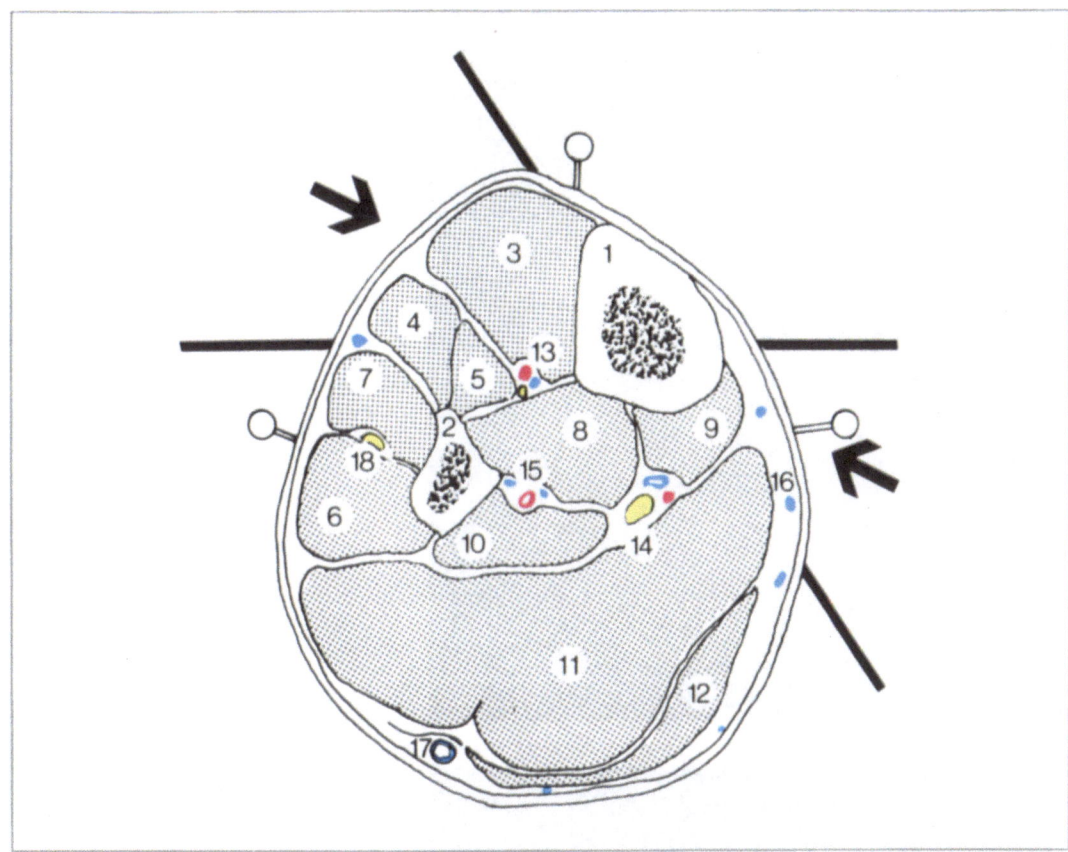

1 Tibia	10 Flexor hallucis longus
2 Fibula	11 Soleus
3 Tibialis anterior	12 Gastrocnemius (medial head)
4 Extensor digitorum longus	13 Anterior tibial vessels + deep peroneal nerve
5 Extensor hallucis longus	14 Posterior tibial vessels + tibial nerve
6 Peroneus longus	15 Peroneal vessels
7 Peroneus brevis	16 Long saphenous vein
8 Tibialis posterior	17 Short saphenous vein
9 Flexor digitorum longus	18 Superficial peroneal nerve

Safe Areas. These are further reduced and lie ventro-laterally and dorso-medially in relation to the tibia.

Cutaneous Zones Related to the Safe Areas. These comprise two zones lying ventro-laterally and dorso-medially in relation to the reference lines.

- *The ventro-lateral zone* consists of the ventral two-thirds of the area between the ventral and lateral reference lines.

- *The dorso-medial zone* lies between the middle of the medial aspect of the tibia and the medial border of the medial head of the gastrocnemius muscle.

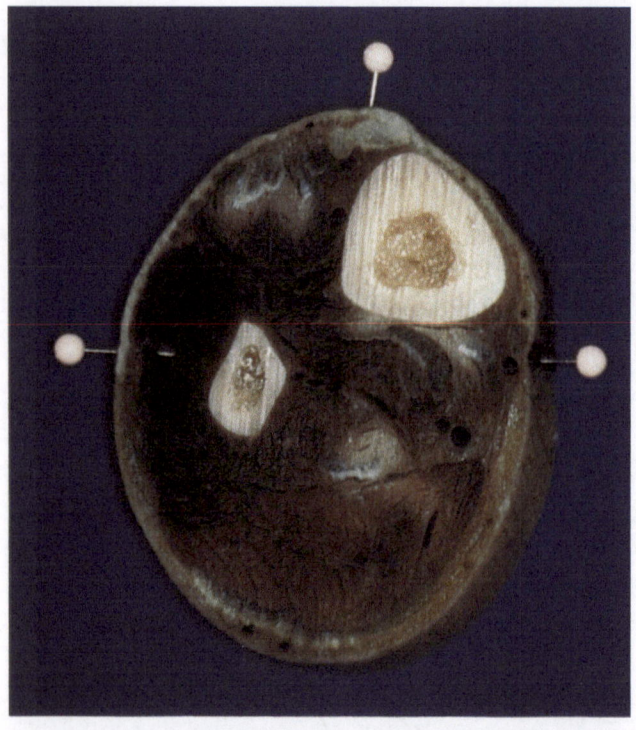

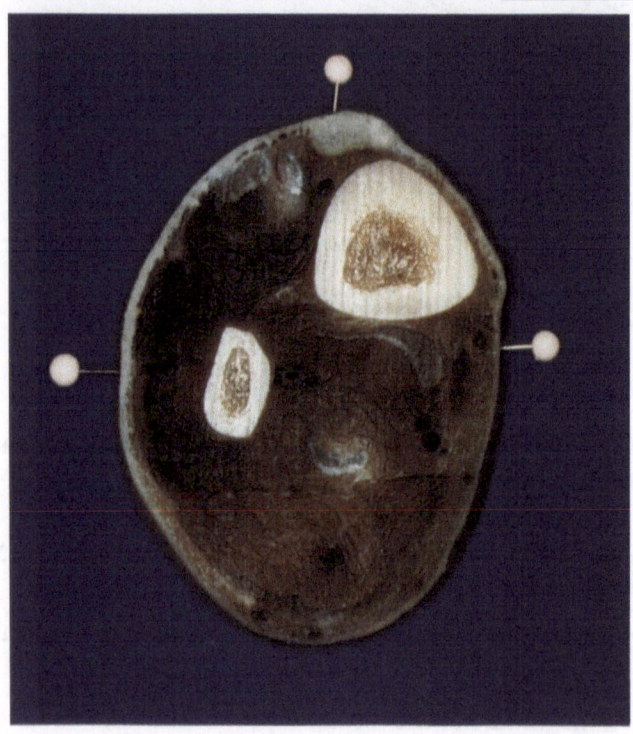

Cross-sections D 8 and D 9

Bones. These two sections have been taken on either side of the junction of the middle and distal thirds of the tibia. The two bones lie at 45° to one another. Section D 8 is proximal to the junction, and Section D 9, distal to it.

Vessels and Nerves. The posterior tibial vessels and the tibial nerve form the largest neurovascular bundle. The remaining vessels and nerves lie further away from the bones.

1 Tibia
2 Fibula
3 Tibialis anterior
4 Extensor digitorum longus
5 Extensor hallucis longus
6 Peroneus longus
7 Peroneus brevis
8 Tibialis posterior
9 Flexor digitorum longus
10 Flexor hallucis longus
11 Soleus
12 Achilles tendon
13 Anterior tibial vessels + deep
     peroneal nerve
14 Posterior tibial vessels + tibial nerve
15 Peroneal vessels
16 Long saphenous vein
17 Short saphenous vein
18 Superficial peroneal nerve

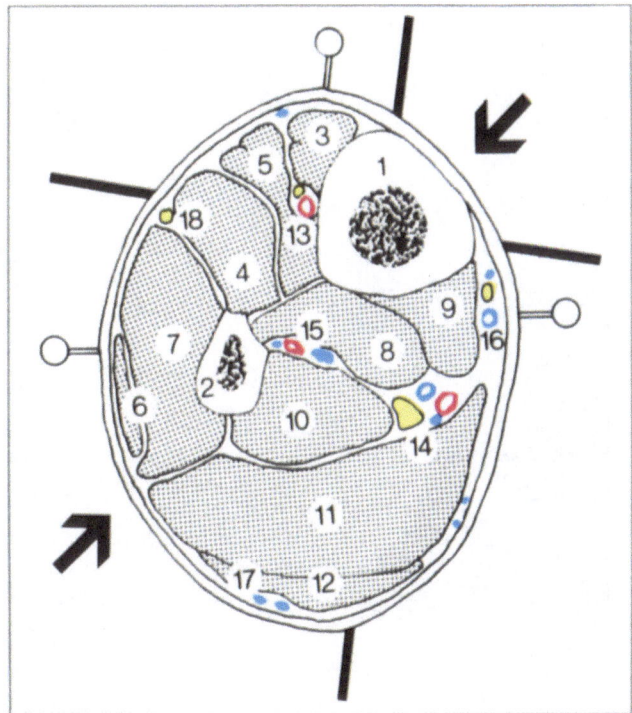

Safe Areas. These are large and lie ventro-laterally and dorso-medially in relation to the tibia.

Cutaneous Zones Related to the Safe Areas. These comprise two zones lying dorso-medially and ventro-medially in relation to the reference lines.

- *The large dorso-medial zone* consists of the area between the ventral border of the body of the peroneus longus muscle and the middle of the triceps surae muscle.

- *The narrow ventro-medial zone* consists of the area between the medial border of the tibialis anterior muscle and the medial border of the tibia.

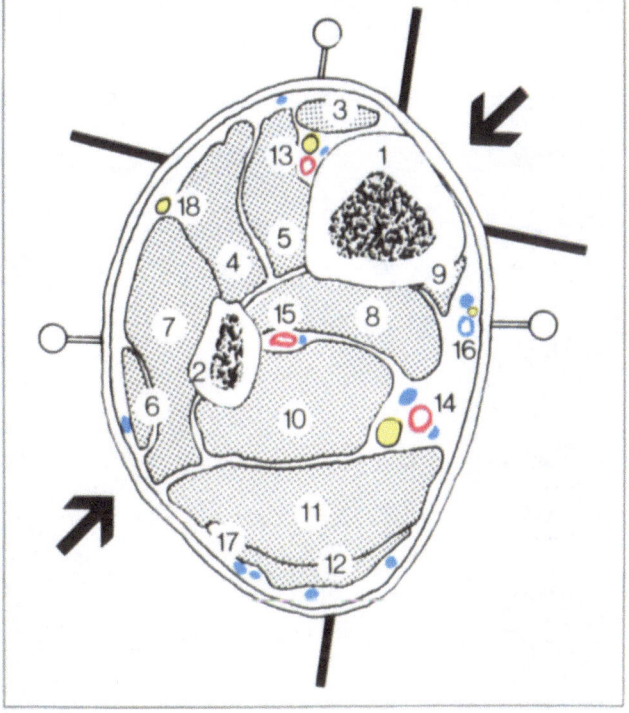

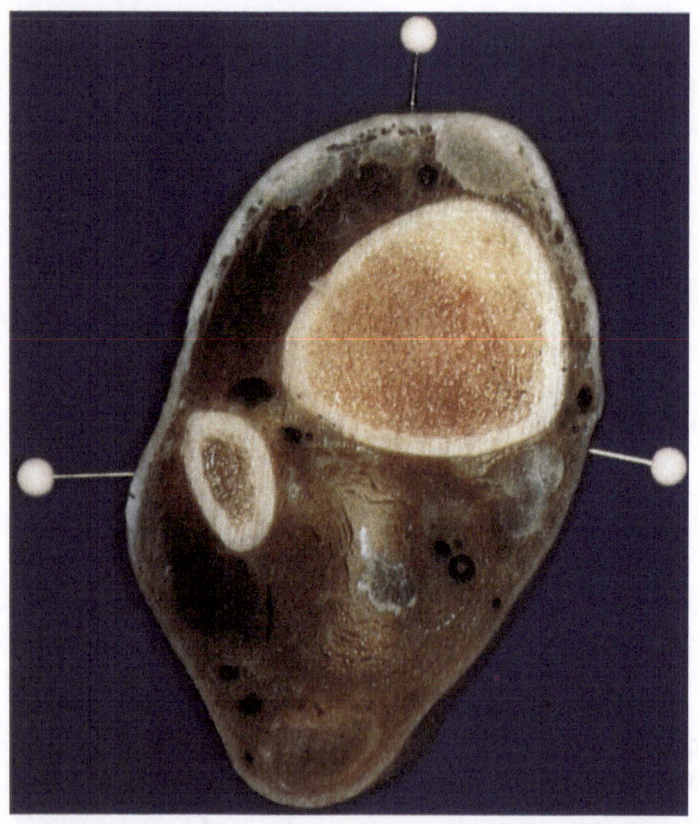

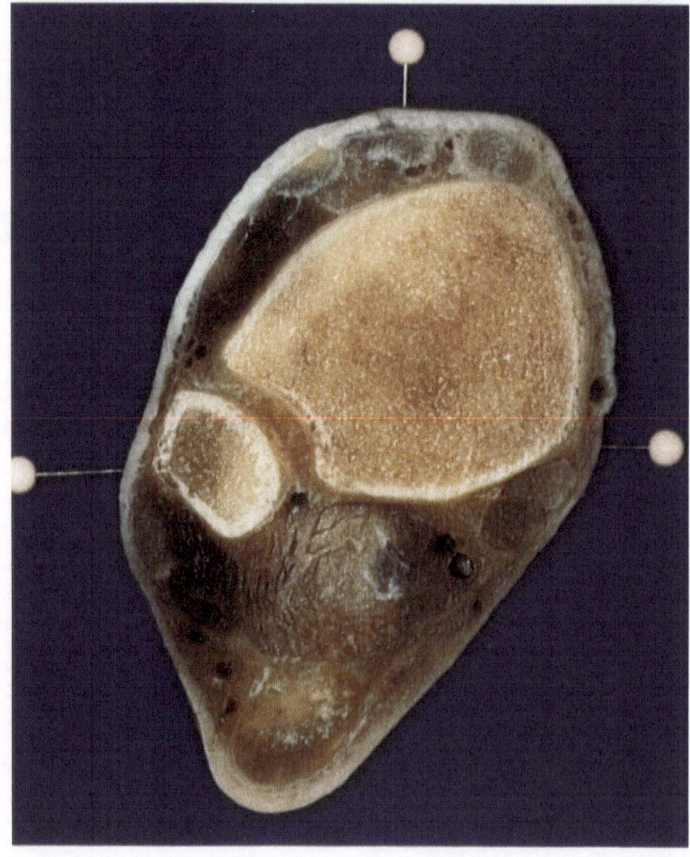

Cross-sections D 10 and D 11

Bones. These two sections show the tibial metaphysis (Section D 10) and epiphysis (Section D 11). The two bones have become larger and lie at 45° to one another.

Vessels and Nerves. The posterior tibial vessels and tibial nerve lie close to the dorsal aspect of the tibia. The anterior tibial vessels are located at the mid-ventral aspect of the tibial epiphysis. The dorsal and ventral tendons should not be transfixed.

1 Tibia
2 Fibula
3 Tibialis anterior
4 Extensor digitorum longus
5 Extensor hallucis longus
6 Peroneus longus
7 Peroneus brevis
8 Peroneus tertius
9 Tibialis posterior
10 Flexor digitorum longus
11 Flexor hallucis longus
12 Triceps surae
   (Achilles tendon)
13 Anterior tibial vessels + deep
   peroneal nerve
14 Posterior tibial vessels +
   tibial nerve
15 Peroneal vessels
16 Long saphenous vein
17 Short saphenous vein

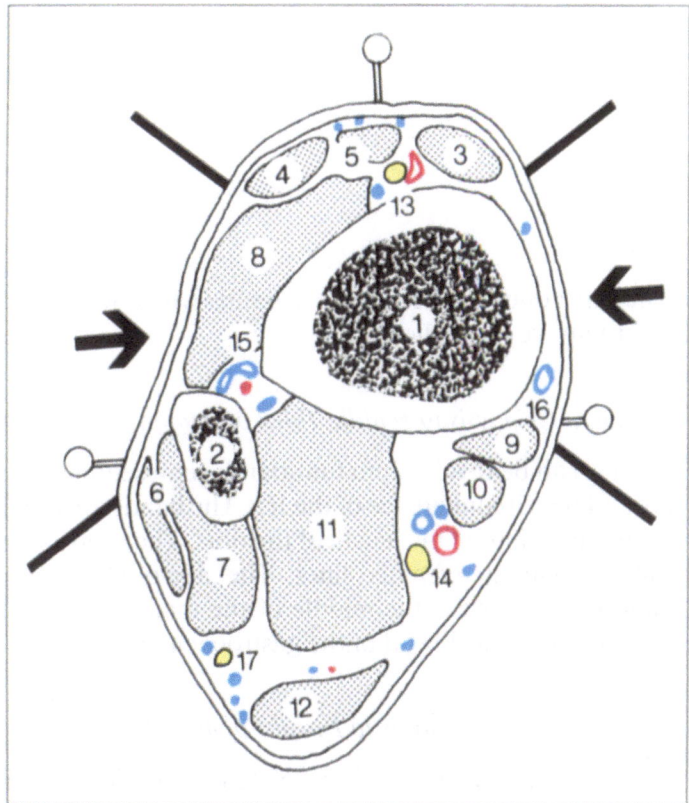

Safe Areas. These are of moderate size and lie laterally and medially in relation to the tibia.

Cutaneous Zones Related to the Safe Areas. These comprise two zones lying ventro-laterally and ventro-medially in relation to the reference lines.

- *The ventro-lateral zone* does not consist of the whole of the area between the ventral and lateral reference lines, but lies between the lateral border of the extensor digitorum longus muscle and the lateral aspect of the fibular metaphysis.

- *The ventro-medial zone* consists of the superficial portion of the distal tibia between the tendons of the tibialis anterior and posterior muscles.

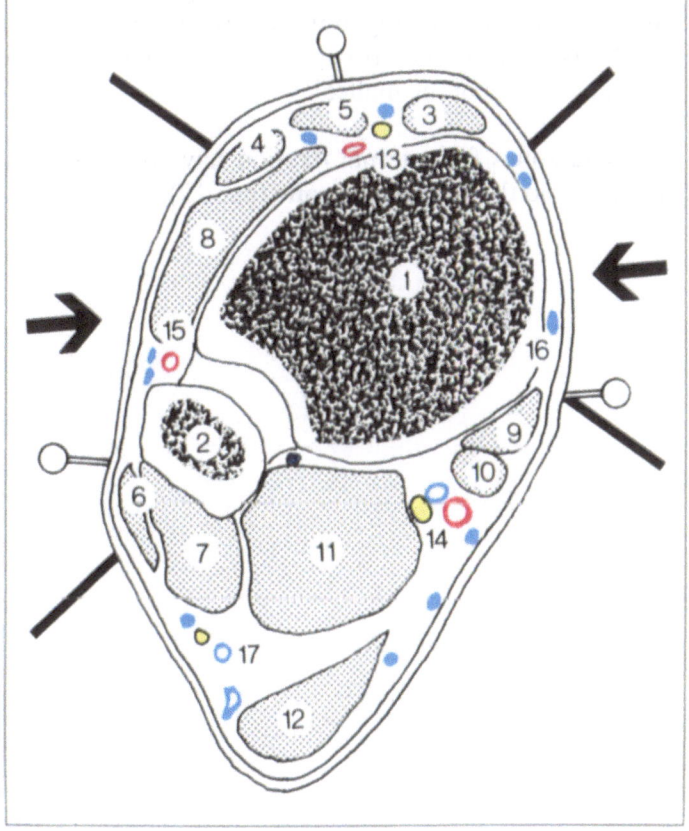

Safe Zones of the Leg

The safe zones can be viewed as bands which wind around the leg from its proximal end to just above the malleoli.

Proximal Epiphysis to Middle Third of Diaphysis (Sections D1–D7)

The leg can be divided into four bands representing alternately safe and prohibited zones around the circumference of the leg. The bands are longitudinal in this region, which represents over half of the length of the tibia. The ventro-lateral band (1) lies between the ventral and lateral reference lines. The dorso-lateral zone (2), diametrically opposed to band 1, lies between the middle of the medial aspect of the tibia and the middle of the body of the medial head of the gastrocnemius muscle.

Distal Third of Diaphysis to Distal Epiphysis (Sections D8–D11)

In this area the two bones lie at 45° to one another. The tibial nerve and the posterior tibial vessels form a substantial bundle, whereas the remaining vessels and nerves have either divided into small branches or are of insignificant size. The direction and size of the bands change at this level owing to these topographical differences. Distal to the middle of the diaphysis, the safe zones become narrower and lie more ventrally. They expand again dorsally at the level of the malleoli.

The safe zones for external fixation have been designated the ventro-lateral safe zone (band 1) and the medial safe zone (band 2).

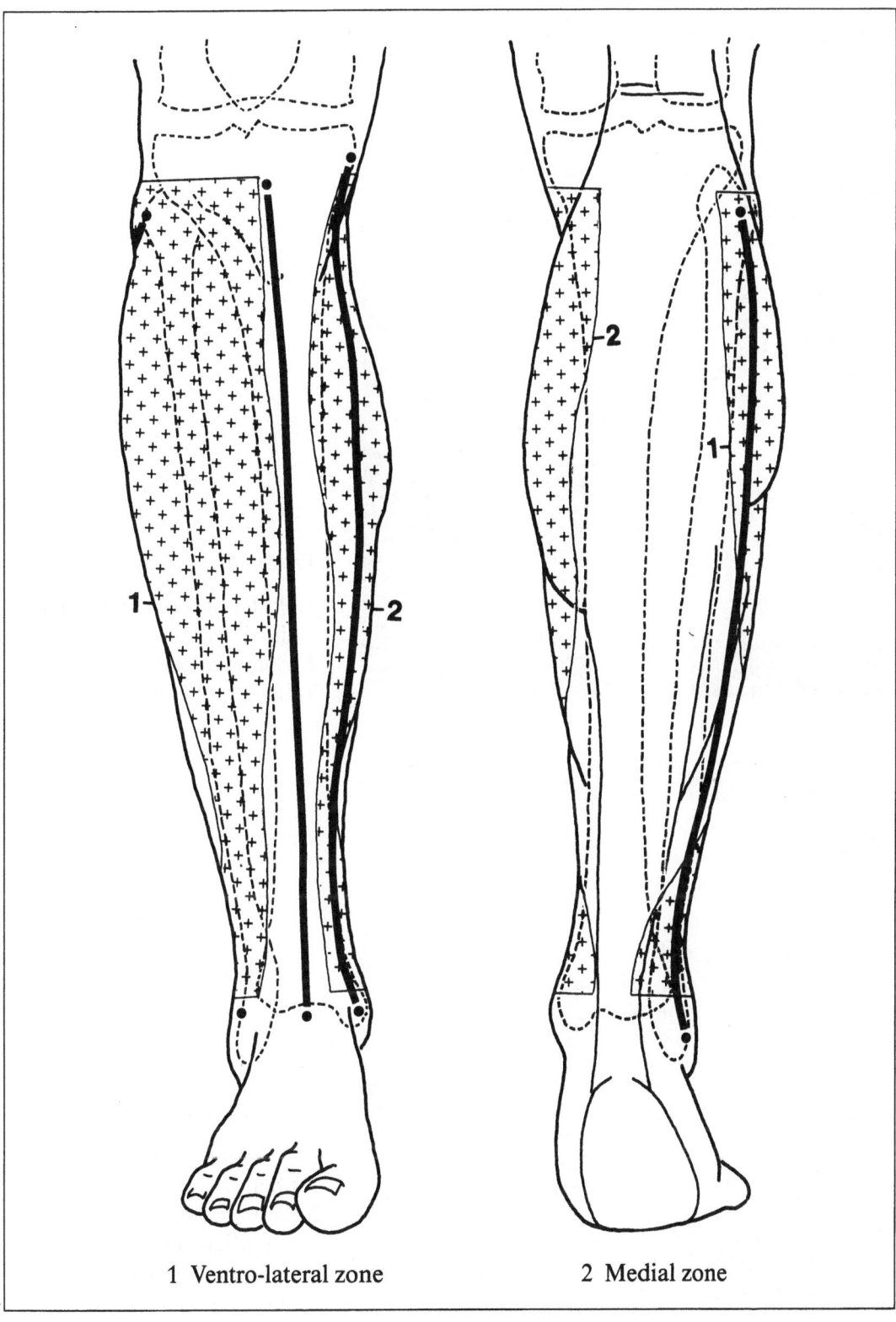

1 Ventro-lateral zone   2 Medial zone

Computerised Tomographic Scans of the Leg

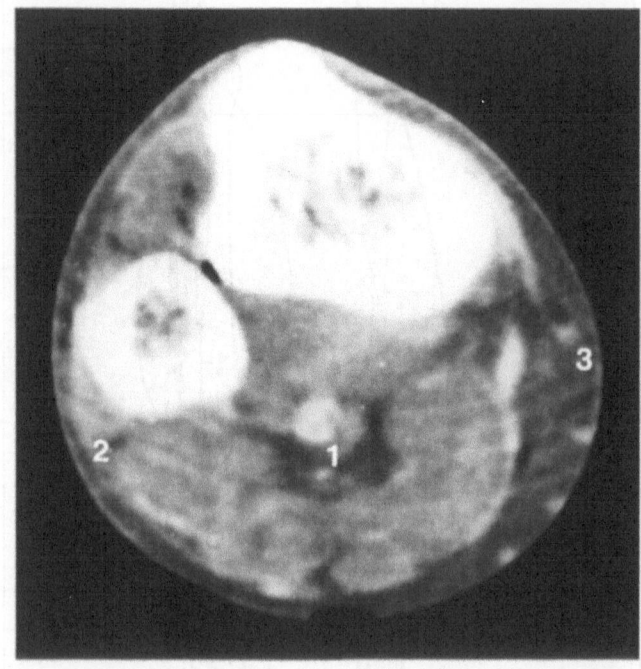

Proximal epiphysis

1 Posterior tibial vessels and
   tibial nerve
2 Common peroneal nerve
3 Long saphenous vein

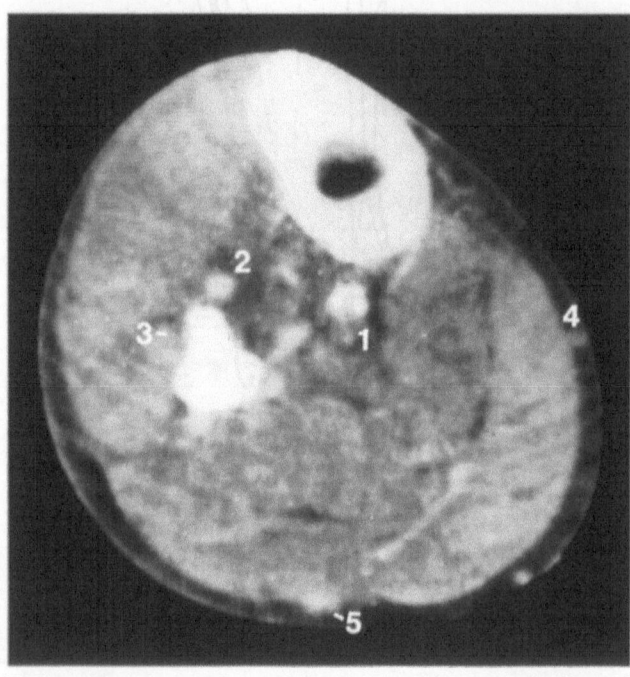

Proximal diaphysis

1 Posterior tibial vessels and
   tibial nerve
2 Anterior tibial vessels
3 Superficial peroneal nerve
4 Long saphenous vein
5 Short saphenous vein

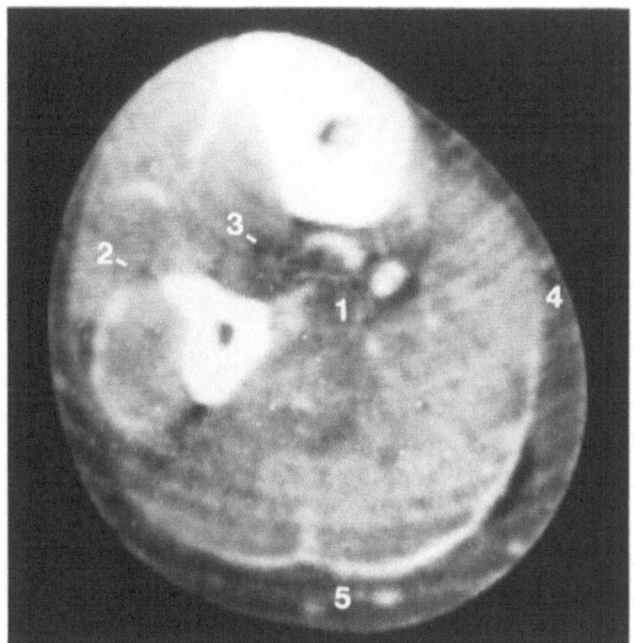

Middle diaphysis

1 Posterior tibial vessels and tibial
   nerve
2 Superficial peroneal nerve
3 Anterior tibial vessels
4 Long saphenous vein
5 Short saphenous vein

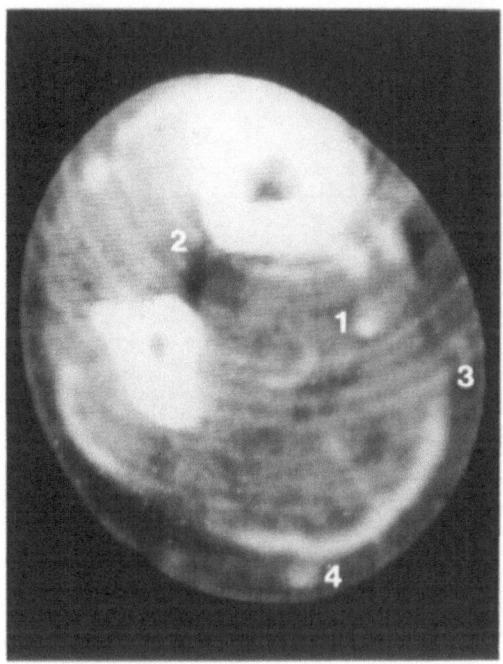

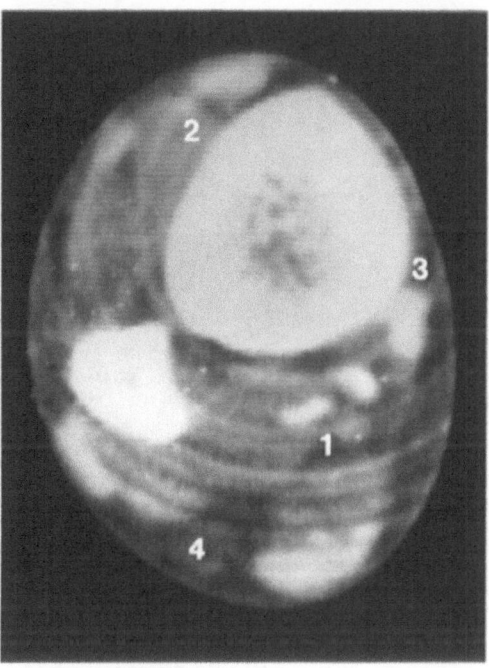

Distal diaphysis

1 Posterior tibial vessels and tibial nerve
2 Anterior tibial vessels
3 Long saphenous vein
4 Short saphenous vein

Distal metaphysis

1 Posterior tibial vessels and tibial nerve
2 Anterior tibial vessels
3 Long saphenous vein
4 Short saphenous vein

Glossary Index

Shoulder and Arm

English terms	International Latin terms	Page number
Bones		
acromion		5
clavicle	clavicula	5
coracoid process	processus coracoideus	6, 7
first rib	costa I	7
humerus		4–24, 26
bicipital groove	sulcus intertubercularis	7, 24
greater tuberosity	tuberculum major	4, 6, 7, 9
head	caput	4, 7, 8
lateral epicondyle	epicondylus lateralis	4, 5, 24
lesser tuberosity	tuberculum minor	4, 6–9, 24, 26
medial epicondyle	epicondylus medialis	4, 5, 22–24
surgical neck	collum chirurgicum	4, 12, 13
trochlea		4, 22
olecranon		4, 22, 23
scapula		7, 9, 11, 13
neck	collum	7
Joints		
acromioclavicular joint	artic.* acromioclavicularis	5
elbow	artic. cubiti	22, 24, 27
scapulo-humeral joint	artic. humeri	6, 7, 9, 10, 24, 26
Muscles		
anconeus	m. anconeus	23
biceps brachii	m. biceps brachii	7, 9, 11, 13, 15–17, 19, 21, 23, 24
long head	caput longum	7, 9, 11, 13, 15, 17, 24
short head	caput breve	9, 11, 15, 17
brachialis	m. brachialis	15–19, 21, 23, 24
brachioradialis	m. brachioradialis	4, 16–19, 21, 23
coracobrachialis	m. coracobrachialis	9, 11, 13
deltoid	m. deltoideus	4, 7, 9, 11, 13, 24
extensor carpi radialis brevis	m. extensor carpi radialis brevis	23
extensor carpi radialis longus	m. extensor carpi radialis longus	21, 23
flexor carpi radialis	m. flexor carpi radialis	23
flexor capi ulnaris	m. flexor carpi ulnaris	23
infraspinatus	m. infraspinatus	9, 11, 13
latissimus dorsi	m. latissimus dorsi	11, 13

Latin	English	Pages
m. pectoralis major	pectoralis major	7, 9, 11, 13
m. pectoralis minor	pectoralis minor	7, 9, 11, 13
m. pronator teres	pronator teres	23
m. scalenus anterior	scalenus anterior	7
m. serratus anterior	serratus anterior	7, 9, 11, 13
m. subclavius	subclavius	7
m. subscapularis	subscapularis	7, 9, 11, 13
m. supraspinatus	supraspinatus	7
m. teres major	teres major	11, 13
m. teres minor	teres minor	9, 11
m. trapezius	trapezius	7, 9, 11
m. triceps brachii	triceps brachii	11, 13, 15, 17, 19, 21
caput laterale	lateral head	13, 15, 17, 19
caput longum	long head	11, 13, 15, 17, 19
caput mediale	medial head	15, 17, 19

Nerves

Latin	English	Pages
n. axillaris	axillary nerve	11, 13
plexus brachialis	brachial plexus	6–10, 26
fasciculus posterior	dorsal cord	6–9
fasciculus lateralis	lateral cord	6–9
fasciculus medialis	medial cord	6–9
n. medianus	median nerve	11, 13–24, 26, 27
n. musculocutaneus	musculocutaneous nerve	13–21, 24, 26, 27
n. radialis	radial nerve	11, 13–24, 26, 27
n. ulnaris	ulnar nerve	11, 13–24, 26, 27

Arteries

Latin	English	Pages
a. axillaris	axillary artery	6–11, 26
a. brachialis	brachial artery	5, 13–24, 26, 27
a. circumflexa humeri posterior	posterior circumflex artery	13
a. profunda brachii	profunda brachii artery	14, 15

Veins

Latin	English	Pages
v. axillaris	Axillary vein	6–11, 26
v. basilica	basilic vein	17, 19, 21, 23
v. brachialis	brachial vein	13–23, 26, 27
v. cephalica	cephalic vein	7, 9, 11, 13, 15, 17, 19, 21, 23
v. circumflexa humeri posterior	posterior circumflex vein	13
V. profunda brachii	profunda brachii vein	14, 15

* artic. = articulatio

Forearm

English terms	International Latin terms	Page number
Bones		
humerus		
lateral epicondyle	epicondylus lateralis	31, 33
medial epicondyle	epicondylus lateralis	31
trochlea		31
pisiform	os pisiformis	33
radius		31
head	caput	30, 31, 33–37, 39–41, 43–49, 51–56, 60, 62, 63
neck	collum	30, 33, 60
styloid process	processus styloideus	30, 34, 60, 62
ulna		31
coronoid process	processus coronoideus	30–56, 58, 60, 62, 63
head	caput	30
olecranon		54, 55
styloid process	processus styloideus	31–33, 58, 62
		31, 54, 55
Joints		
distal radio-ulnar joint	artic. radioulnaris distalis	30, 52, 54, 56
proximal radio-ulnar joint	artic. radioulnaris proximalis	32
radiocarpal joint	artic. radiocarpea	30, 54, 56
Muscles		
abductor pollicis longus	m. abductor pollicis longus	21, 45, 49, 53, 55
anconeus	m. anconeus	33
biceps brachii	m. biceps brachii	32, 33, 35
brachialis	m. brachialis	33, 35
brachioradialis	m. brachioradialis	33–35, 37, 41, 45, 47–49, 51, 53
extensor carpi radialis brevis	m. extensor carpi radialis brevis	33, 37, 38, 41, 45, 49, 53, 55
extensor carpi radialis longus	m. extensor carpi radialis longus	33–35, 37, 41, 45, 49, 53, 55
extensor carpi ulnaris	m. extensor carpi ulnaris	33, 35, 37, 41, 45, 49, 51, 53, 55
extensor digiti minimi	m. extensor digiti minimi	35, 37, 41, 45, 49, 53, 55
extensor digitorum communis	m. extensor digitorum communis	33, 35, 37, 41, 42, 45, 49, 53, 55
extensor indicis	m. extensor indicis	45, 49, 53, 55
extensor pollicis brevis	m. extensor pollicis brevis	45, 49, 53, 55

Thigh

Leg

English terms	International Latin terms	Page Number
Bones		
fibula		97–105, 109, 111, 113, 115, 116
head	caput	97–99
lateral malleolus	malleolus lateralis	96, 97, 116
neck	collum	100, 101
tibia		96–116
medial malleolus	malleolus medialis	96, 97, 116
tuberosity	tuberositas	96, 97, 99, 101–103
Joints		
patellar ligament	lig. patellae	99
proximal tibio-fibular joint	artic. fibularis proximalis	96, 98–100
tibial collateral ligament	lig. collaterale tibiale	99, 101, 103
Muscles		
extensor digitorum longus	m. extensor digitorum longus	99–101, 103, 107, 109, 111, 113, 115
extensor hallucis longus	m. extensor hallucis longus	109, 111, 113, 115
flexor digitorum longus	m. flexor digitorum longus	105, 107, 109, 111, 113, 115
flexor hallucis longus	m. flexor hallucis longus	105, 107, 109, 111, 115
gastrocnemius	m. gastrocnemius	99, 101, 103, 105, 107, 109, 111, 116
lateral head	caput laterale	99, 101, 103, 105, 107, 109, 111, 116
medial head	caput mediale	99, 101, 103, 105, 107, 109, 111, 116
gracilis	m. gracilis	99, 101, 103
peroneus brevis	m. peroneus brevis	108–111, 113, 115
peroneus longus	m. peroneus longus	101, 103–105, 107–111, 113, 115
peroneus tertius	m. peroneus tertius	115
popliteus	m. popliteus	99, 101, 103
sartorius	m. sartorius	99, 101, 103
semitendinosus	m. semitendinosus	99, 101
soleus	m. soleus	99–101, 103, 105, 107, 109, 111, 113, 115
tibialis anterior	m. tibialis anterior	99–101, 103, 105, 107, 109, 111, 113, 115
tibialis posterior	m. tibialis posterior	103, 105, 109, 111, 113, 115
triceps surae	m. triceps surae	113, 115
Achilles tendon	Achillis tendo	113, 115

B. G. Weber, F. Magerl, St. Gallen, Switzerland

The External Fixator

AO/ASIF-Threaded Rod System
Spine-Fixator
With a Chapter by C. Brunner
Foreword by A. Sarmiento
Translated from the German by T. Telger
1985. 362 partly colored figures. XV, 373 pages.
ISBN 3-540-134756-4

Contents: Preamble. – The ASIF Threaded External Fixator in General Orthopaedics and Trauma Surgery of the Extremities: General Part. On the Biomechanics of External Fixation. The Threaded External Fixator. Instrumentation. Operative Technique for the Threaded External Fixator. – Techniques for Reinforcing the External Fixator. – Situations Requiring a Special Operative Technique. Axial Corrections. Local Care Following Frame Application. Duration of External Fixation, and Removal of the External Fixator. Instrumentation Used with the Threaded External Fixator. The Threaded External Fixator in Adults. Clinical Examples. The Threaded External Fixator in Children and Adolescents. Concluding Remarks. – External Spinal Skeletal Fixation. – References. – Subject Index.

Manual on the AO/ASIF Tubular External Fixator

By **G. Hierholzer,** Duisburg, Germany; **I. Rüedi,** Chur; **M. Allgöwer,** Bern, Switzerland; **J. Schatzker,** Toronto, Canada
1985. 104 figures, some in colour. V, 100 pages.
ISBN 3-540-13518-9

Contents: Introduction and Basic Indications for the Use of External Skeletal Fixation. – Mechanical Principles of External Skeletal Fixation. – Remarks Concerning the Pathophysiology of Compound Fractures. – Indications for External Skeletal Fixation Versus Internal Fixation. – Four Building Components of the AO Tubular System and the Accompanying Surgical Instruments. – Basic Assemblies and Their Use. – Technical Details for Construction. – Clinical Application of External Skeletal Fixator. – Appendix: Special Indications for the Tubular External Fixator. – Addendum. – References. – Subject Index.

Springer-Verlag
Berlin Heidelberg New York
London Paris Tokyo

A. E. Freeland, M. E. Jabaley, J. L. Hughes, University of Mississippi, Jackson, USA

Stable Fixation of the Hand and Wrist

1986. 960 figures, 4 tables. XIII, 285 pages.
ISBN 3-540-96300-6

Contents: History and Basic Science. – Fracture Repair: Metacarpals and Carpals. Phalangeal Fractures. Distal Radial Fractures. Special Fracture Categories. – Reconstruction: Arthrodesis. Corrective Osteotomies. Other Corrective Osteotomies. Other Reconstruction. – Epilogue. – Index.

F. Sequin, R. Texhammar, Bern, Switzerland

AO/ASIF Instrumentation

Manual of Use and Care

Introduction and Scientific Aspects by H. Willenegger
Translated from the German by T. Telger
1981. Approx. 1300 figures, 17 separated Checklists.
XVI, 306 pages. ISBN 3-540-10337-6

Contents: Introduction. – Medical and Scientific Directives. – Principles of the AO/ASIF-Technique and Basic Mechanical Principles. – Practical Part: Instrumentation of the AO/ASIF. Compressed Air and Compressed-Air Machines. Cleaning, Care, and Sterilization of Instruments and Implants. Preoperative, Operative, and Postoperative Guidelines. Suggestions for the Management of Various Fractures. Preparation of the Instruments. – Subject Index.

J. Guyot, University of Besançon, France

Atlas of Human Limb Joints

Illustrations by J. L. Vannson

Springer-Verlag
Berlin Heidelberg New York
London Paris Tokyo

Translated from the French by R. A. Elson
1981. 113 figures. X, 252 pages. ISBN 3-540-10380-5